Technology and Regulation

2021
Volume 3

OPEN PRESS TiU

TECHNOLOGY AND REGULATION 2021

Volume 3

DOI: 10.26116/techreg.volume.2021
ISBN: 9789403670621

Technology and Regulation
Tilburg Institute for Law, Technology, and Society (TILT)
Tilburg Law School
P.O. Box 90153
5000 LE Tilburg
The Netherlands
techreg.org

Principal Contact:
Ronald Leenes
Editor-in-Chief
Tilburg Institute for Law, Technology, and Society (TILT), Tilburg Law School
r.e.leenes@tilburguniversity.edu

Support Contact:
Aaron Martin
a.k.martin@uvt.nl

Published by: Open Press TiU
Contact details: info@openpresstiu.edu
https://www.openpresstiu.org/

Cover Design by: Claudia Tofan & KaftWerk/Janine Hendriks

Open Press TiU is the academic Open Access publishing house for Tilburg University and beyond. As part of the Open Science Action Plan of Tilburg University, Open Press TiU aims to accelerate Open Access in scholarly book publishing.

OPEN PRESS Tilburg University 2022

Aims and Scope

Technology and Regulation (TechReg) is an international journal of law, technology and society, with an interdisciplinary identity. TechReg provides an online platform for disseminating original research on the legal and regulatory challenges posed by existing and emerging technologies (and their applications) including, but by no means limited to, the Internet and digital technology, artificial intelligence and machine learning, robotics, neurotechnology, nanotechnology, biotechnology, energy and climate change technology, and health and food technology. We conceive of regulation broadly to encompass ways of dealing with, ordering and understanding technologies and their consequences, such as through legal regulation, competition, social norms and standards, and technology design (or in Lessig's terms: law, market, norms and architecture). We aim to address critical and sometimes controversial questions such as: How do new technologies shape society both positively and negatively? Should technology development be steered towards societal goals, and if so, which goals and how? What are the benefits and dangers of regulating human behaviour through technology? What is the most appropriate response to technological innovation, in general or in particular cases? It is in this sense that TechReg is intrinsically interdisciplinary: we believe that legal and regulatory debates on technology are inextricable from societal, political and economic concerns, and that therefore technology regulation requires a multidisciplinary, integrated approach. Through a combination of monodisciplinary, multidisciplinary and interdisciplinary articles, the journal aims to contribute to an integrated vision of law, technology and society. We invite original, well-researched and methodologically rigorous submissions from academics and practitioners, including policy makers, on a wide range of research areas such as privacy and data protection, security, surveillance, cybercrime, intellectual property, innovation, competition, governance, risk, ethics, media and data studies, and others.

TechReg is double-blind peer-reviewed and completely open access for both authors and readers. TechReg does not charge article processing fees.

Editorial Team

CONTENTS

01

Technology and Regulation

Keeping up with cryptocurrencies

How financial regulators used radical innovation to bolster agency reputation

Lauren Fahy*, Scott Douglas** & Judith van Erp***

Bureaucratic reputation, reputation management, financial regulation, regulatory agencies, innovation, cryptocurrency

l.a.fahy@uu.nl

s.c.douglas@uu.nl

j.g.vanerp@uu.nl

Invented in 2008 with Bitcoin, cryptocurrencies represent a radical technological innovation in finance and banking; one which threatened to disrupt the existing regulatory regimes governing those sectors. This article examines, from a reputation management perspective, how regulatory agencies framed their response. Through a content analysis, we compare communications from financial conduct regulators in the UK, US, and Australia. Despite the risks, challenges, and uncertainties involved in cryptocurrency supervision, we find regulators treat the technology as an opportunity to bolster their reputation in the immediate wake of the Global Financial Crisis. Regulators frame their response to cryptocurrencies in ways which reinforce the agency's ingenuity and societal importance. We discuss differences in framing between agencies, illustrating how historical, political, and legal differences between regulators can shape their responses to radical innovations.

1. Introduction

The financial sector is experiencing a wave of radical innovation unmatched since the popular adoption of the Internet. Innovation can drive economic growth and better quality of life.[1] Yet, its disruptive nature poses challenges for regulators.[2] Cryptocurrencies are a case in point. Emerging in 2008, cryptocurrencies like Bitcoin have brought new types of technically complex and ever-evolving products into financial markets. Cryptocurrencies exacerbated risks financial regulators typically supervise and introduced new risks. Cryptocurrencies work very differently to traditional forms of currency, payment, and money transfer. It was not immediately clear whether their use was legal, and whether it should be.[3] How do regulatory agencies respond to this kind of radical innovation?[4]

Legal and regulatory governance scholarship often focuses its analysis of this question, fittingly, on legal and operational responses. These are the ways regulators reform rules and practices to continue to efficiently manage market risks e.g. revising regulations. There is a rich literature describing, analysing, and evaluating such responses.[5] Prior studies, however, also show a 'political' dimension to how regulators respond. Different stakeholders have different economic interests in, and ideological positions on, how innovation will be regulated.[6] Regulators are sensitive to these tensions. They want to build stakeholder support for, or at least avoid criticism about, their legal and operational responses.[7] Agencies may do so through choosing legal/operational responses which are broadly acceptable to the public.[8] They may also try to maintain/build stakeholder support through strategic communications about those responses.[9] Research,

1 Cristie Ford, *Innovation and the State: Finance, Regulation, and Justice* (Cambridge University Press 2017) 7.

2 Ford (n 1) 16–17.

3 Douglas W Arner, Janos Barberis and Ross P Buckley, 'The Evolution of FinTech: A New Post-Crisis Paradigm' (2015) 47 *Georgetown Journal of International Law* 1271.

4 Radical innovations, here, are inventions which significantly reduce the costs of key inputs in a way that significantly transforms sectors, economies, or societies (as opposed to gradual, 'incremental' innovations) (C. Freeman and L. Soete, *The Economics of Industrial Revolution (London: Pinter 1997))*. Cryptocurrencies, and the underlying technology of blockchain, have the potential to reduce the costs of financial products and services and are proving disruptive to financial markets, as well as adjacent markets like financial law and accounting (Ford ((n1)) 49; Arner et al. ((n2)) 7).

5 e.g. R Brownsword, E Scotford and E Yeung, 'Law, Regulation, and Technology: The Field, Frame, and Focal Questions' in R Brownsword, E Scotford and Yeung (eds), *The Oxford Handbook of Law, Regulation and Technology* (Oxford University Press 2017); Karen Yeung, 'How Is the UK Responding to the Technologies of the Fourth Industrial Revolution?' [2017] Ethics, Law, & Society 102; Gregory N Mandel, 'Emerging Technology Governance', *Innovative governance models for emerging technologies* (Edward Elgar 2013).

6 ML Jones and J Millar, 'Hacking Metaphors in the Anticipatory Governance of Emerging Technology: The Case of Regulating Robots', *The Oxford Handbook of Law, Regulation, and Technology* (Oxford University Press 2017).

7 Moshe Maor, 'Organizational Reputation and Jurisdictional Claims: The Case of the U.S. Food and Drug Administration' (2010) 23 Governance 133.

8 Maor (n 7).

9 Amit Tzur, 'Uber Über Regulation? Regulatory Change Following the Emergence of New Technologies in the Taxi Market' (2019) 13 Regulation & Governance 340; EF Gerding, *Law, Bubbles, and Financial Regulation (Routledge 2016)*; M Lee, *'The Legal Institutionalization of Public Participation in the EU Governance of Technology', The Oxford handbook of law, regulation, and technology (Oxford University Press 2017).*

* Lauren Fahy is a PhD fellow at Utrecht School of Governance, Utrecht University, the Netherlands.

** Scott Douglas is assistant professor of Public Management at Utrecht School of Governance, Utrecht University, the Netherlands.

*** Judith van Erp is professor of Regulatory Governance at Utrecht School of Governance, Utrecht University, the Netherlands.

Received 24 Dec 2020, Accepted 17 Mar 2021, Published 1 Apr 2021.

however, has not yet systematically and empirically analysed the kinds of communication strategies agencies use, and why.

Reputational theory has been increasingly applied to analyse political dimensions of regulatory agency behaviour.[10] Reputation is the image of the agency held in the minds of its audiences (e.g. the public, politicians, companies). Reputation is what those audiences imagine the agency to be like; "a set of symbolic beliefs about the unique or separable capacities, roles, and obligations of an organization, where these beliefs are embedded in audience networks".[11] Reputational theories argue that, when faced with a new problem or task, agencies will consider how their response will be perceived. In responding, they seek to manage their reputation so that they maintain audience support.[12] Agencies manage their reputation in various ways, including 'symbolic' strategies; through the use of public relations, communications, and marketing.[13]

How, though, do regulatory agencies symbolically manage their reputation in response to the specific challenges posed by radical technological innovation? To answer this question, we draw primarily on bureaucratic reputation theory.[14] This theory provides a framework to describe and compare the symbolic strategies agencies use[15] and explain why agencies choose some strategies over others.[16] Bureaucratic reputation thus provides a strong basis to analyse agency reputation management in the face of new kinds of regulatory challenge. The unique features of innovation governance as a regulatory task are little discussed in theory and rarely empirically examined.[17] This study aims to begin to address this gap.

In this study, we compare communications about cryptocurrencies from three financial conduct regulators in the United Kingdom, United States, and Australia. We use quantitative and qualitative content analysis to determine what kind of symbolic reputation management strategies these agencies used. We then apply a bureaucratic reputation theoretical framework to draw out possible explanations as to why regulators chose the responses they did, analysing responses in historical, political, and legal context.

This study contributes to theory by presenting a more comprehensive framework for describing and explaining how regulatory agencies manage reputation in the face of radical innovation. Through the case study, we illustrate how such a framework helps us understand the political dimension of regulator responses to innovation. The study illuminates that reputational considerations can deter regulators from intervening to govern radical innovations. Under certain circumstances, however, and — as the cryptocurrency case shows — a desire to bolster agency reputation can actually drive regulators to involve themselves in even the most risky, uncertain, and challenging radical innovations.

2. Case background

Cryptocurrencies began with Bitcoin. In 2008, Satoshi Nakaomoto (a pseudonym for a group of individuals) released Bitcoin's open-source code. Alongside, Nakaomoto published a paper. It argued that, in the Internet age, relying on financial institutions to pay one another was inefficient and risky. Bitcoin would eliminate the need.[18] Cryptocurrencies are systems by which to send and receive payments through an encryption system run on a decentralized network of computers. They allow users to pay one another through digital transfers in (more or less) real time, like cash, and without mediation by a bank or any third party.[19]

Today cryptocurrencies have become more mainstream and commercial. Some people use cryptocurrencies as originally intended: as an online payment system. Others buy cryptocurrencies as an investment or as speculation. Some uses of cryptocurrencies – or uses in some jurisdictions –are illegal, some legally ambiguous, and some fully legal (for example, the regulated Gemini exchange in New York).[20] We can now understand cryptocurrencies as part of a large wave of radical innovation in finance in the post-Global Financial Crisis period (along with the rise of other 'fintech' like crowdfunding and financial AI). We are still in the midst of this wave, which is introducing new kinds of businesses, products, and ideas to the market.[21]

This study, however, is concerned with how regulators respond to radical innovations as they emerge. Our analysis looks to the first decade after cryptocurrencies were invented. Our case study focuses on three financial conduct regulators: the New York State Department of Financial Services (NY DFS), the Financial Conduct Authority of United Kingdom (UK FCA), and the Australian Securities and Investments Commission (AUS ASIC). These regulators began to publicly acknowledge cryptocurrency trading in their jurisdictions around 2012. At that time, cryptocurrencies were a strange, fringe development. As cryptocurrencies were different to existing financial technologies, they fell outside many legal definitions such as 'currency', 'financial institution', and 'derivative'.[22] Governments, regulators, and courts were still determining how they should be defined and regulated. Such questions were legally complex, and difficult to answer given the novelty and technical complexity of cryptocurrencies.[23] Regulatory agencies had to consider whether and how to intervene on cryptocurrencies given (typically) gaps in policy and law. Cryptocurrencies, however, were also a controversial topic, of interest to con-

10 Jan Boon, Heidi H Salomonsen and Koen Verhoest, 'A Reputation for What, to Whom, and in Which Task Environment: A Commentary' [Forthcoming] Regulation & Governance.
11 Daniel Carpenter, *Reputation and Power: Organizational Image and Pharmaceutical Regulation at the FDA (Princeton University Press 2010) 45.*
12 Moshe Maor, 'Theorizing Bureaucratic Reputation' in A Waeraas and Maor, Moshe (eds), *Organizational Reputation in the Public Sector (Routledge 2015).*
13 Carpenter (n 11) 70.
14 Daniel Carpenter, *The Forging of Bureaucratic Autonomy: Reputations, Networks, and Policy Innovation in Executive Agencies, 1862-1928* (Princeton University Press 2001); (n 11).
15 Sharon Gilad and T Yogev, 'How Reputation Regulates Regulators: Illustrations from the Regulation of Retail Finance', *Oxford Handbook of Corporate Reputation (Oxford University Press 2012); Saar Alon-Barkat, 'Can Government Public Communications Elicit Undue Trust? Exploring the Interaction between Symbols and Substantive Information in Communications' (2020) 30 Journal of Public Administration Research and Theory 77; Dovil Rimkut , 'Organizational Reputation and Risk Regulation: The Effect of Reputational Threats on Agency Scientific Outputs' (2018) 96 Public Administration 70.*
16 Daniel Carpenter and George A Krause, 'Transactional Authority and Bureaucratic Politics' (2015) 25 Journal of Public Administration Research and Theory 5; Moshe Maor, Sharon Gilad and Pazit Ben-Nun Bloom, 'Organizational Reputation, Regulatory Talk, and Strategic Silence' (2013) 23 Journal of Public Administration Research and Theory 581.
17 Maor (n 7).
18 Joshua Davis, 'The Crypto-Currency' (*The New Yorker, 3 October 2011)* https://www.newyorker.com/magazine/2011/10/10/the-crypto-currency (accessed 21 December 2020).
19 A Narayan and others, *Bitcoin and Cryptocurrency Technologies: A Comprehensive Introduction (Princeton University Press 2016) ix–xxiii.*
20 Nate Lanxon and Olga Kharif, 'Winklevoss Twins' Crypto Exchange Is Expanding Into the U.K.' *Bloomberg.com (24 September 2020)* https://www.bloomberg.com/news/articles/2020-09-24/winklevoss-twins-crypto-exchange-is-expanding-into-the-u-k (accessed 21 December 2020).
21 Arner, Barberis and Buckley (n 3).
22 Ford (n 1) 143.
23 Davis (n 18).

sumers, politicians, and business.[24] As the next section outlines, we would expect regulators under these circumstances to manage their reputation very carefully as they respond to this radical innovation.

3. Theoretical framework

3.1 Radical innovation: A reputational threat to be managed?

How do regulatory agencies symbolically manage their reputation in the face of innovation in their jurisdiction? Presently, bureaucratic reputation theory provides a partial answer. Two studies to date have examined the field of innovation governance.[25] Both examined the US Food and Drug Administration's response to innovation in the pharmaceutical sector.

In his study, Maor developed a model applying bureaucratic reputation theory to explain regulatory responses to radical innovation. Specifically: to explain and predict when agencies will and will not claim their legal authority extends over novel technologies. Claims, here, can refers to statements which explicitly or implicitly demonstrate the agency believes it has authority e.g. policy statements, issuing guidelines.[26]

When deciding how to respond to innovation, Maor argues, regulators do not simply consider objective, technical and legal questions (e.g. does our current legal authority cover this new biotechnology?). They will also consider how their response will be perceived by their audiences.[27] How will their response affect the agency's reputation? In bureaucratic reputation theory, a strong reputation is one of an agency's most important assets. A reputation is strong when most people in a group (or many groups across society) like, or at least accept the legitimate existence of, that organization.[28] A strong reputation helps agencies to survive and achieve their goals. A weak reputation makes agencies less effective, and at risk from having their funding cut, or being eliminated altogether.[29] Agencies are thus highly motivated to manage the reputation. They want to influence audience perceptions in ways that maintain or build support for the agency and its actions (rather than eliciting public questioning, criticism, or defiance).[30]

Regulators make decisions about responding to innovation in this context.[31] Maor contests that regulators are risk averse: they prioritize minimizing anticipated reputational damage over pursuing opportunities.[32] Regulators prefer to pursue the low hanging fruit of easy regulatory wins over tackling unwieldy problems.[33] Radically new technologies are uncertain, hard to regulate, and controversial.[34] Jurisdictional claims over novel technologies can fail.[35] Even if regulators gain authority to act, their responses are likely to be deemed a failure in whole or in part due the complexities of supervision and mixed public opinion about what constitutes success.

To minimize risks, agencies prefer to delay making claims over novel technologies (or never make them at all).[36] Regulators want time to consider and/or prepare a solid claim. They also want time to build a coalition of supporters for that claim. Agencies have different kinds of audiences who could form such a coalition (politicians, business, consumers etc.). Agencies want to build and maintain support with as many audiences as possible, especially those audiences critical to their survival and success.[37] Different audiences, though, often have different interests, ideologies, and preferences. It thus takes time for agencies to secure support from various audiences to make a claim.

While agencies prefer to (indefinitely) delay their response to innovation, this strategy can become untenable. Delaying a claim can do more damage to the agency's reputation if certain, other 'threats' arise. One such threat is negative publicity. New information may be published showing this novel technology is harmful e.g. this unregulated medical practice is killing people. Agency audiences then start criticizing the agency for its negligence. Negative publicity makes agencies more likely to make a timely claim.[38] Other bureaucratic reputation research reinforces negative public attention increases the likelihood of a quick response.[39][40]

The second category of threat driving claims concerns how *other* regulatory agencies respond. Novel technologies tend to potentially fall under the authority of two or more agencies. This can incentivize regulators to make a claim quickly before others can.[41] Agencies want to avoid a scenario where other agencies make competing claims over technologies they themselves want to supervise.[42] Competition can damage their relationship with professional colleagues.[43] Further, agencies typically do not want to risk having to share authority.[44] They do not want to share authority over specific technologies nor the broader regulatory field.[45] Sharing responsibilities means regulators have less autonomy; leaving them open to criticism about a technology whose supervision they cannot fully control.[46] Sharing or losing authority like this can, too, make the regulator come to be seen as less *unique*.

Agencies, ideally, want to build and then maintain a *unique* reputa-

24 Davis (n 18).
25 Maor (n 7); Carpenter (n 11).
26 Maor (n 7) 134.
27 Maor (n 7) 134.
28 Carpenter (n 11) 45.
29 Carpenter (n 11) 727.
30 Carpenter (n 11) 752–3.
31 Maor (n 7) 134.
32 Maor (n 7) 138; see also: RK Weaver, 'The Politics of Blame Avoidance' (1986) 6 Journal of Public Policy 371; Christopher Hood, *The Blame Game: Spin, Bureaucracy, and Self-Preservation in Government* (Princeton University Press 2011); Judith van Erp, 'New Governance of Corporate Cybersecurity: A Case Study of the Petrochemical Industry in the Port of Rotterdam' (2017) 68 Crime, Law and Social Change 75.
33 Keith Hawkins, *Environment and Enforcement: Regulation and the Social Definition of Pollution* (Oxford University Press 1984) https://oxford.universitypressscholarship.com/view/10.1093/acprof:oso/9780198275145.001.0001/acprof-9780198275145 (last accessed 21 December 2020).
34 Ford (n 1); S Ranchordás, 'On Sharing and Quasi-Sharing: The Tension between Sharing-Economy Practices, Public Policy, and Regulation', *The rise of the sharing economy: Exploring the challenges and opportunities of collaborative consumption* (Praeger 2018).
35 Maor (n 7) 137.
36 Maor (n 7) 137.
37 Maor, Gilad and Bloom (n 16) 583; Sharon Gilad, Saar Alon Barkat and Alexander Braverman, 'Large-Scale Social Protest: A Business Risk and a Bureaucratic Opportunity' (2016) 29 Governance 371.
38 Maor (n 7) 139.
39 In bureaucratic reputation theory, responses can be either in the form of communicating, like issuing a press releases, or substantive action, like increasing regulatory resources to address a risk.
40 Maor, Gilad and Bloom (n 16); Carpenter and Krause (n 16).
41 Maor (n 7) 140.
42 see also: JQ Wilson, *Bureaucracy: What Government Agencies Do and Why They Do It* (Basic Books 1989); Madalina Busuioc, 'Friend or Foe? Inter-Agency Cooperation, Organizational Reputation, and Turf' (2016) 94 Public Administration 40.
43 Maor (n 7) 141.
44 Busuioc (n 42).
45 For example, if a second agency claims authority over one biotechnology this may give them a foothold to claim authority over the supervision of medical technologies in general.
46 Wilson (n 42); Busuioc (n 42).

tion. They want to be seen as the sole provider of a public good or service in their jurisdiction. Agencies seen to make a unique contribution are more recognized, socially valued, and harder for politicians to attack or replace.[47] In the case of innovation, agencies are more likely to make a quick claim if they think it will build their unique reputation.*48* Conversely, agencies are less likely to make claims over technologies peripheral to their unique reputation. This reflects a more general tendency for agency reputation management to be path-dependent.[49] Once agencies establish their unique position in their society — one which elicits support from enough audiences — they tend to seek to maintain rather than change that reputation.[50] Maor argues, in the case of innovation, unusual claims over areas traditionally regulated by someone else upsets the business community. That audience wants agencies to stick to "traditional goals and areas of oversight, rather than innovative forms...".[51] One possible exception is if the agency who should be traditionally responsible does not make the obvious claim. A 'vacuum' can lead to more negative publicity, compelling the regulator to respond.[52]

Maor explored the validity of this model through an analysis of actual claims by the Food and Drug Administration over biotechnologies.[53] His analysis supports the expectations discussed thus far. This would imply that, when faced with innovation, regulators prefer *not* to respond or take responsibility. This argument is broadly supported by findings from scholarship on innovation law and governance.[54] A major limitation of such accounts, however, is they assume regulators always see innovation as a threat.

3.2 Expanding the framework: Innovation as a reputational opportunity

In the main, bureaucratic reputation scholarship examines agency reputation management in cases where, either: 1) events are inherently threats e.g. crises, scandals[55] or 2) agencies are theorized to perceive them as threats.[56] In his theoretical model, Maor maps these assumptions onto the field of innovation governance. Yet, we cannot assume, a priori, regulators see innovation in these terms.

Carpenter's[57] research shows agencies do not always respond to external events purely as threats. Agencies are not always risk-averse. They can recognize externals events, like innovation, as opportunities to strengthen reputation. Agencies do not simply react to negative publicity to fulfil audience demands. Rather, agencies have some capacity to: 1) frame how audiences perceive external events and the agency's response to them, and 2) choose who their audiences are. Agencies can use language and symbolism to shape how the public understands the opportunities and risks of an event, and court support from new and different audiences.[58]

Carpenter theorizes more directly about technological innovation in his 2010 study of the US Food and Drug Administration. Carpenter's study shows innovation can be a reputational opportunity for regulators, first, because it creates opportunities for agencies to build their unique reputation. New technologies mean new kinds of public goods and 'bads' (i.e. regulatory risks to be managed).[59] This creates opportunities for agencies to do something new and of societal value. Second, innovation can introduce new audiences for an agency and shift the relative power of audiences (e.g. with the influx of different kinds of businesses to a market).[60] In his study, the Food and Drug Administration proactively cultivated support for the agency and its interventions into the development of new pharmaceuticals. They did so through their practical actions, but also through their communications: through the use of discourse, rhetoric, language, and symbolism.[61]

Combining Maor and Carpenter's perspectives provides a more nuanced and realistic picture of how regulatory agencies manage their reputation in the face of innovation. Yet, neither author systematically examines what symbolic reputation management strategies agencies use and why. Further, both perspectives were developed through studies of the same regulator, in the same sector, in the same country. It is not clear how well this extends to other contexts.[62] This study builds upon theoretical frameworks to date, and provides an analytical framework to describe and explain symbolic reputation management in the face of innovation. Further, we explore the validity of this framework through a case study in a significantly different context (finance in the US, UK, and Australia).

3.3 Analytical framework

Another strand of bureaucratic reputation research provides us with the basis for our analytical framework.[63] This research has catalogued the kinds of symbolic reputation management strategies agencies use. Critical to this theory is that agency reputation is multi-dimensional. Audiences judge agencies on several different kinds of criteria. This study draws upon the criteria Carpenter[64] proposes: how well the agency delivers quality outputs and outcomes (*performative reputation*); how expert the agency is (*technical reputation*), how well it follows required or desirable processes (*procedural reputation*), and how ethical and good its goals and means are (*moral reputation*).[65]

47 Carpenter (n 11) 45.
48 Maor (n 7) 140.
49 Maor (n 12) 25; Wilson (n 42) 76.
50 Sharon Gilad, 'Political Pressures, Organizational Identity, and Attention to Tasks: Illustrations from Pre-Crisis Financial Regulation' (2015) 93 Public Administration 593; Arjen Boin and others, 'Does Organizational Adaptation Really Matter? How Mission Change Affects the Survival of U.S. Federal Independent Agencies, 1933–2011' (2017) 30 Governance 663.
51 Maor (n 7) 140.
52 Maor (n 7) 141.
53 Maor (n 7).
54 Erik F Gerding, 'Code, Crash, and Open Source: The Outsourcing of Financial Regulation to Risk Models and the Global Financial Crisis' (2009) 84 Washington Law Review 127(n 9); Ford (n 1) 48; Rob Frieden, 'Adjusting the Horizontal and Vertical in Telecommunications Regulation: A Comparison of the Traditional and a New Layered Approach' (2003) 55 55 Federal Communications Law Journal 207 (2003) https://www.repository.law.indiana.edu/fclj/vol55/iss2/3; RG Lee and J Petts, 'Adaptive Governance for Responsible Innovation', *Responsible Innovation: Managing the Responsible Emergence of Science and Innovation in Society* (Wiley 2013).
55 e.g. Moshe Maor and Raanan Sulitzeanu Kenan, 'The Effect of Salient Reputational Threats on the Pace of FDA Enforcement' (2013) 26 Governance 31.
56 George A Krause and J Kevin Corder, 'Explaining Bureaucratic Optimism: Theory and Evidence from U.S. Executive Agency Macroeconomic Forecasts' (2007) 101 The American Political Science Review 129.
57 (n 14).
58 Carpenter (n 14) e.g. 144; 234-244; 310.
59 see also: Busuioc (n 42).
60 Carpenter (n 11) 72; see also: Kevin Young, 'Financial Industry Groups' Adaptation to the Post-Crisis Regulatory Environment: Changing Approaches to the Policy Cycle' (2013) 7 Regulation & Governance 460.
61 Carpenter (n 11) e.g. 60; 66-67.
62 Boon, Salomonsen and Verhoest (n 10).
63 Rimkuté (n 15); Madalina Busuioc and Dovilé Rimkuté, 'The Promise of Bureaucratic Reputation Approaches for the EU Regulatory State' (2020) 27 Journal of European Public Policy 1256; Gilad and Yogev (n 15); Alon-Barkat (n 15).
64 (n 11).
65 Carpenter (n 11) 45–46.

Table 1. Carpenter's conceptual framework of agency reputation

Competency	Description
Performative	Concerns agency outputs i.e. how well they are doing the task at hand or achieving their goals.
Moral	Concerns the normative aspects of the agency i.e. the moral value of its goals or its behaviors (e.g. demonstrating compassion).
Technical	Concerns the extent to which the agency has necessary expertise in relevant areas.
Procedural	Concerns how well the agency follows required or desirable processes e.g. administrative, legal.

In their communications, agencies try to shape how audiences perceive them and their actions.[66] They use language and symbols designed to 'signal' to audiences that they are, for example, an ethical organization whose actions are based on technical expert judgements. In this study, we refer to this behaviour as 'image management strategy'.[67] Agencies may frame themselves or their actions with more emphasis on some dimensions of reputation over others.[68] Agencies will also emphasize more specific 'aspects' within dimensions. For example, while selling itself on good moral reputation, one agency might discuss the aspect of protecting consumers while another might focus on facilitating market competition.[69]

Agencies further try to shape how audiences perceive them through making strategic choices about whether to communicate in a high- or low- profile manner (here: 'communications strategy'). Agencies sometimes choose a strategy of 'positive visibility'.[70] They communicate a lot and in forums designed to attract public attention. Alternatively, agencies may be 'strategically silent', communicate very little, and/or in forums designed to have a smaller audience.[71] In the context of responding to innovation, agencies also make strategic choices about image management. Centrally: whether they should frame their response as consistent with their existing image, or a departure from that image.[72]

Which strategies, then, would we expect regulators to choose when faced with innovation? As presented in the theoretical framework, this depends on what the agency is like, what the innovation is like, how audiences perceive the innovation and the agency, and how other agencies respond. These factors are summarized in Figure 1. Prior to a detailed analysis of the cases, we cannot make specific predictions as to which strategies each agency will choose. Our aim is not to develop universal "singular laws"[73] for how regulators manage reputation in the face of innovation. Rather, in the following analysis of the cryptocurrency case, we aim to illustrate how applying a reputational lens — and this framework in particular — to innovation governance can help scholars better understand how and why regulators respond as they do.

4. Methodology

We chose cryptocurrency as an extreme case of innovation.[74] As will be discussed further, cryptocurrencies are a case of *radical innovation*.[75] Cryptocurrencies represent a substantial departure from previous technologies, rather than an incremental improvement.[76] Radical innovations are especially challenging – technically and politically – for regulators to manage.[77] Extreme cases are useful for exploratory research; to probe – in this case – how agencies respond and the possible reasons for those responses in an "open-ended fashion".[78]

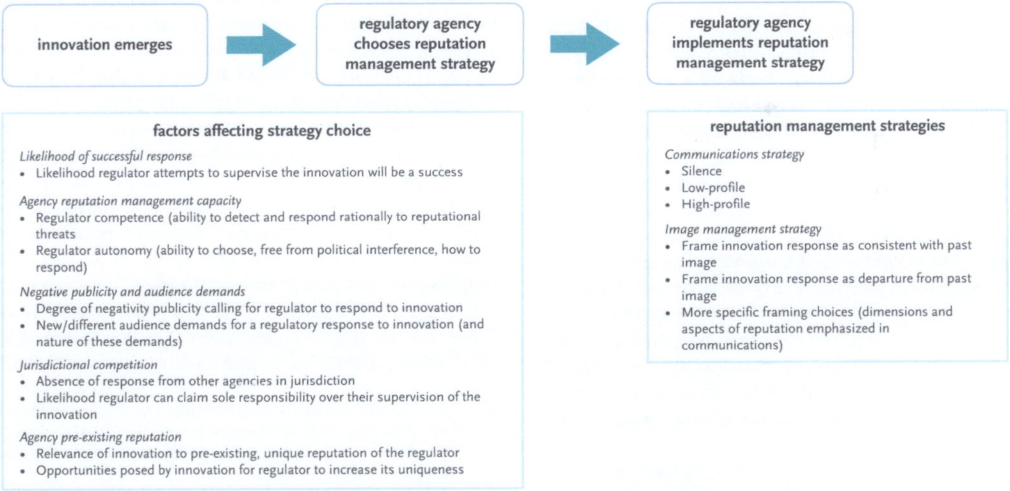

Figure 1. Regulatory agency symbolic reputation management in the face of innovation: Theoretical framework

66 Carpenter (n 11) 70; Manuela Moschella and Luca Pinto, 'Central Banks' Communication as Reputation Management: How the Fed Talks under Uncertainty' (2019) 97 Public Administration 513.
67 Arild Wæraas and Haldor Byrkjeflot, 'Public Sector Organizations and Reputation Management: Five Problems' (2012) 15 International Public Management Journal 186, 190.
68 Rimkuté (n 15); Gilad and Yogev (n 15); Tom Christensen and Åse Gornitzka, 'Reputation Management in Public Agencies: The Relevance of Time, Sector, Audience, and Tasks' (2019) 51 Administration & Society 885.
69 Wæraas and Byrkjeflot (n 67) 190.
70 Gilad, Alon Barkat and Braverman (n 37).
71 Maor, Gilad and Bloom (n 16).
72 Gilad and Yogev (n 15); Maor and Sulitzeanu Kenan (n 55); Carpenter (n 11) 68; Rimkuté (n 15) 6.
73 Carpenter (n 11) 754.
74 Jason Seawright and John Gerring, 'Case Selection Techniques in Case Study Research: A Menu of Qualitative and Quantitative Options' (2008) 61 Political Research Quarterly 294, 301.
75 Ford (n 1) 49.
76 Kevin Zheng Zhou, Chi Kin (Bennett) Yim and David K Tse, 'The Effects of Strategic Orientations on Technology- and Market-Based Breakthrough Innovations' (2005) 69 Journal of Marketing 42.
77 Brownsword, Scotford and Yeung (n 5).
78 Seawright and Gerring (n 74) 302.

In this study we compare reputation management responses of three regulators (NY DFS, UK FCA, and AUS ASIC). We sought to compare a manageable number of cases which were from broadly similar contexts: Anglophone, OECD liberal democracies with large, well-established financial markets and rapidly growing fintech sectors.[79] We chose agencies, too, which were similar. All three agencies included are financial conduct regulators, with responsibilities including consumer protection, with formal autonomy from government.[80] We examined which communication strategy each agency chose and whether, and how, they engaged in image management. Image management was determined through comparing the image they presented in their communications about cryptocurrency to their image in the period immediately prior, then comparing between cases. The before and after, and inter-agency, comparisons increases our confidence agencies chose particular strategies in response to cryptocurrency trading.

The study used three methods: 1) qualitative document review of the agency's pre-existing image and 2) quantitative and 3) qualitative content analysis of cryptocurrency communications. The quantitative analysis determined communications strategy. The document analysis, with the qualitative content analysis, analysed image management.

For the document analysis, we searched Google Scholar, Westlaw, and Lexis Nexis with agency titles, acronyms, and 'reputation'. Documents were included if they were published in the three years prior to the agency's first communication about cryptocurrency. Documents included the agency's own statements, academic literature, and authoritative media and expert judgements. To determine the nature of the agency's pre-existing image, documents were interpreted using the coding schema described below.

For the quantitative content analysis, we collected all agency communications published after 2008 and before March 2018 about cryptocurrency (a total of 538 individual texts). These were imported into NVIVO and analysed to determine text type and audience.[81] Agencies were considered to have chosen low- or high- profile strategy based on number of texts, frequency of publishing, and high- versus low-profile fora (e.g. targeted, private speeches versus media appearances). A sample of 351 texts were then subjected to qualitative content analysis to determine what kind of image each agency presented. We developed a coding schema using Carpenter's framework of reputational competencies and informed by previous analyses using that framework.[82] The schema was applied to determine what overall image agencies were signalling.[83] This was then compared with the competencies and aspects, presented by the other two agencies, and compared to its pre-existing image. In the final stage, we compared the images agencies presented with their pre-existing reputation, and with the reputation presented by the other two cases.

5. Findings and analysis

In this section, we first present findings of the quantitative and qualitative content analysis. We then move on to an interpretive analysis. We apply our theoretical framework to draw out some historical, political, and legal case factors which help to explain why regulators responded in this way, and why we see some differences between reputation management by different agencies.

5.1 Findings of the content analysis

5.1.1 Low- or high- profile communications strategy?

The quantitative content analysis found all three regulators chose a high-profile communications strategy. Agencies published texts about cryptocurrencies frequently. Figure 2 shows regulators consistently communicate on the topic. Agencies display somewhat different preferences for specific text types (e.g. speeches versus mass media). Yet, the most common text types were those one would usually use to target mass audiences: tweets, press releases, and web pages (Figure 3). Thus, agencies can be said to have responded to cryptocurrencies in ways one would expect to draw public attention.

5.1.2 (How) do agencies engage in image management?

This section discusses each regulator's image prior to cryptocurrency trading (results of the document analysis) and whether and what signals were different in cryptocurrency communications (results of the qualitative content analysis).

NY DFS

The New York State Department of Financial Services was founded in 2011 in response to the perceived failure of previous regulatory arrangements to prevent the Global Financial Crisis. Perhaps as a result, NY DFS emphasized moral competencies first and foremost. The agency presented itself as a consumer protector standing up to Wall Street to ensure fair play. Performatively, the regulator portrayed itself as tough, strong, and unyielding. As having "worked aggressively to protect consumers, prevent systematic risk and encourage financial services to thrive and create jobs" [84]. The regulator characterized a prominent enforcement action against a large bank as protecting the United States against "terrorists, weapons dealers, drug kingpins and corrupt sectors".[85] Early enforcement successes led the press to characterize NY DFS as performatively "muscular", [86] and "the new cop"[87]. Superintendent Ben Lawsky was profiled as "Wall Street's Sheriff[88]; a "marathon-running lawyer" with a "taste for

79 Z/Yen, 'The Global Financial Centres Index - Long Finance' (2018) https://www.longfinance.net/programmes/financial-centre-futures/global-financial-centres-index/ (last accessed 22 December 2020; EY, 'EY FinTech Adoption Index 2017: The Rapid Emergence of Fintech' https://www.ey.com/en_kw/financial-services--emeia-insights/the-rapid-emergence-of-fintech (accessed 20 December 2020).

80 On this basis, we chose a US state regulator over a federal agency. US financial regulation is heavily decentralized, partially because the US market is so large (Brian Knight, Federalism and Federalization on the Fintech Frontier, 20 VAND. J. ENT. & TECH. L. 129 (2017)). In mandate and market size, NY DFS is more comparable to UK FCA and AUS ASIC than a federal regulator like the Securities and Exchange Commission.

81 Moschella and Pinto (n 66) 520.

82 e.g. Rimkuté (n 15), described in detailed at Appendix 1.

83 Hsiu-Fang Hsieh and Sarah E Shannon, 'Three Approaches to Qualitative Content Analysis' (2005) 15 Qualitative Health Research 1277, 124–5.

84 NY DFS, 'DFS Annual Reports | Department of Financial Services' (*2011 First Annual Report of the Superintendent to the Governor and Legislature*, 2012) 6 https://www.dfs.ny.gov/reports_and_publications/dfs_annual_reports (last accessed 23 December 2020).

85 cited in Justin O'Brien and Olivia Dixon, 'The Common Link in Failures and Scandals at the World's Leading Banks' (2013) 36 Seattle University Law Review 941, 960.

86 Liz Rappaport, 'Wall Street's New Watcher' *Wall Street Journal* (3 October 2011) https://online.wsj.com/article/SB10001424052970203405504576605790712611496.html (accessed 23 December 2020).

87 Danny Hakim, 'Expanding Reach, Cuomo Creates Second Cop on Financial Beat (Published 2012)' *The New York Times* (29 January 2012) https://www.nytimes.com/2012/01/30/nyregion/financial-services-agencys-reach-spurs-criticism-of-cuomo.html (last accessed 23 December 2020).

88 Jessica Silver-Greenberg and Ben Protess, 'Benjamin Lawsky, Sheriff of Wall Street, Is Taking Off His Badge (Published 2015)' *The New York Times* (20 May 2015) https://www.nytimes.com/2015/05/21/business/dealbook/benjamin-lawsky-to-step-down-as-new-yorks-top-financial-regu-

Wall Street blood". [89] Procedurally, NY DFS presented itself as willing to 'go rogue' in the pursuit of its objectives, even overriding norms of inter-regulator coordination.[90]. In its cryptocurrency communications, NY DFS shows little attempt at manage its image away from this reputation.

NY DFS framed cryptocurrencies as a new area of supervisory activity in which they had obvious jurisdiction.

> 'If there was money transmission going on [in cryptocurrency trading] as the state regulatory in New York we had a very specific regulatory obligation to license those entities, examine those entities, and otherwise regulate those entities in New York'.[91]

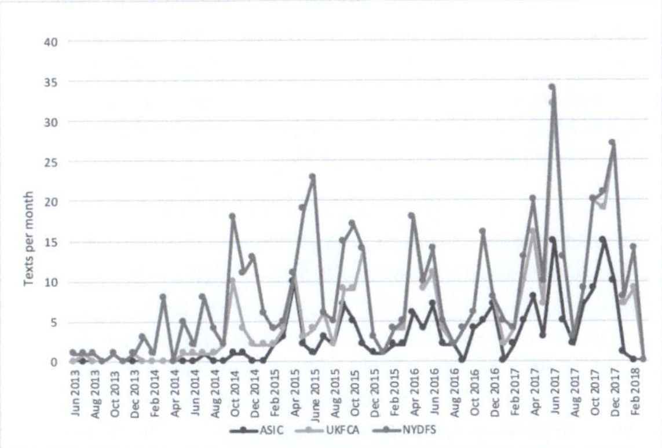

Figure 2. Relevant texts published by regulator over time

Table 3. Texts by type (as percentage)

Text type	AUS ASIC	UK FCA	NY DFS
Tweet	35.1%	46.4%	56.8%
Web page	16.8%	14.1%	1.4%
Speech	11.5%	16.1%	3.6%
Press release	14.1%	8.9%	14.4%
Mass media	2.1%	2.1%	13.7%
Other	20.3%	12.4%	9.6%
total	100%	100%	100%

In discussing the quality of the agency's involvement in cryptocurrency, NY DFS emphasized the moral, performative, and procedural competencies consistent with its established image. The agency presented itself as the same tough regulator, intervening to take on cryptocurrency supervision to protect consumers and combat illegal activity.

> 'If virtual currencies remain a virtual Wild West for narcotraffickers and other criminals, that would not only threaten our country's national security, but also the very existence of the virtual currency industry as a legitimate business enterprise...It is vital to put in place appropriate safeguards for consumers and law abiding citizens'.[92]

Also consistent with its pre-existing image, NY DFS suggests its performance on cryptocurrency regulation cannot and should not be undermined by federal regulation. The agency argues state-based regulators are more experienced than federal, and especially more experienced with regulating non-bank financial entities.[93]

> 'DFS has proven that the state regulatory system is the best way to supervise and cultivate a thriving fintech industry, like virtual currency'.[94]

Some signals NY DFS sent in cryptocurrency communications, however, were different. First, NY DFS emphasized the performative uniqueness and novelty of its approach to cryptocurrency in ways not previously seen. In August 2015, NY DFS introduced the BitLicense scheme. Any firm seeking to use cryptocurrency for finance or banking purposes had to obtain a 'BitLicense' in order to operate legally.[95] The agency repeatedly emphasized they were the first in the nation (and the world) to implement this kind of system.

> 'NY DFS proposed a first-in-the-nation, comprehensive regulatory framework for firms dealing in virtual currency, including Bitcoin'.[96]

Second, NY DFS framed its involvement not only in terms of enforcement but also facilitation. Indeed, the agency positions themselves morally as aiming to enabling financial innovation generally.

> '...We also want to make sure that we don't clip the wings of a fledgling technology before it gets off the ground. We want to make certain that New York remains a hub for innovation and a magnet for new technology firms'.[97]

Performatively, the agency argued it was already regulating in ways which either did not hurt, or indirectly helped, business.

> 'Numerous fintech companies have already succeeded and grown under this regulatory framework...In implementing regulations for the licensing and supervision of virtual currency entities, DFS enhanced trust and legitimacy of a promising emerging financial services technology'.[98]

Third, and finally, signals about NY DFS's procedural competencies have a different emphasis in discussions of cryptocurrency supervision. Whereas the agency had previous presented itself as willing to

lator.html (accessed 23 December 2020).

89 Simon Neville, 'Ben Lawsky: Marathon Man Who Became the Latest Scourge of Wall Street' (*the Guardian*, 11 August 2012) http://www.theguardian.com/business/2012/aug/12/benjamin-lawsky-profile (last accessed 23 December 2020).

90 Jill Treanor, 'Standard Chartered Chief Says Bank Does Not Need to Change Culture' (*the Guardian*, 8 August 2012) http://www.theguardian.com/business/2012/aug/08/standard-chartered-chief-defends-bank (accessed 23 December 2020).

91 NY DFS, 'Superintendent Lawsky Issues Notice of Intent to Hold Public Hearing Regarding Virtual Currencies on January 28 and 29 in New York City.' 1 https://www.dfs.ny.gov/about/press2013/memo1308121.pdf.

92 Ben Lawsky, 'Notice of Inquiry on Virtual Currencies' 1 https://dfs.ny.gov/about/press2013/memo1308121.pdf.

93 Maria Vullo, 'Superintendent's Letter Comptroller's Licensing Manual Draft Supplement: Evaluating Charter Applications from Financial Technology Companies. Letter from Maria Vullo to the Honourable Thomas J. Curry,'.

94 Vullo (n 92) 2.

95 'New York's Bitcoin Hub Dreams Fade with Licensing Backlog' (*CNBC*, 31 October 2016) https://www.cnbc.com/2016/10/31/new-york-bitcoin-hub-dreams-fade-with-licensing-backlog.html (last accessed 23 December 2020).

96 NY DFS, '2014 Annual Report of the New York State Department of Financial Services' 6 https://www.dfs.ny.gov/system/files/documents/2020/03/dfs_annualrpt_2014.pdf.

97 NY DFS, 'Superintendent Lawsky Issues Notice of Intent to Hold Public Hearing Regarding Virtual Currencies on January 28 and 29 in New York City.' (n 90) 2.

98 Vullo (n 92) 6.

violate procedural norms to get results, on cryptocurrency NY DFS signals it is making decisions on cryptocurrency based on rigorous inquiry and fact-finding.

Notably, in the NY DFS case and in regard to the other two regulators, technical competencies were not significantly emphasized. NY DFS does make occasional reference to having general experience in regulating the New York financial market, and once or twice to lacking expertise on cryptocurrencies (discussed further below).

UK FCA

Like NY DFS, the United Kingdom's Financial Conduct Authority was established to replace a regulator implicated in the Crisis (the Financial Services Authority).[99] The UK FCA similarly emphasized its moral, performative, and procedural competencies in the period preceding cryptocurrency trading. Morally, UK FCA presented a renewed moral mission and standards of behaviour. Procedurally, it emphasized ongoing commitment to accountability and transparency while avoiding rigid, rule-based supervision.[100] Performatively, the regulator emphasized the quality of its approach, rather than the strength of its regulation. In particular, that its approach was proactive, responsive, outcome-focused, and suitably flexible. The UK FCA described itself as having performative characteristics of "curiosity", being "already on the case", and demonstrating "professional excellence". [101] The UK FCA liked to characterize itself as leading the world in creative solutions.[102] Further, that the regulator was morally committed to, and performatively demonstrated, a balance in promoting competition and protecting consumers.[103] In communicating about cryptocurrency, UK FCA presented a largely similar image.

Formally, the UK FCA has argued that, until or unless the use of cryptocurrencies constitutes a financial product, they do not have the necessary powers to regulate. [104][105] In their communications, however, UK FCA placed cryptocurrency and fintech supervision generally front and centre in their regulatory brand.[106] The regulator has argued, indeed, that their statutory obligations compel them to take a role.

> 'So, our duty to promote competition is actually, it's full title is 'competition in the interests of consumers'. So, you know that's where we start [our approach to fintech] from'.[107]

In characterizing the agency's approach to cryptocurrencies, UK FCA continued to send strong performative and moral signals that it was a principles-based, outcomes-focused, flexible, and proactive regulator.

> 'In addition to supporting individual businesses, we look to add more flexibility to our regulatory framework and identify barriers to entry for innovative firms...Our approach is typically to regulate the outcome, rather than the specific process'.[108]

Perhaps in this spirit, the UK FCA launched 'Project Innovate' in 2014. Project Innovate was composed of an Innovation Hub[109] and regulatory sandbox. The sandbox allowed new kinds of fintech including cryptocurrency and related technology to be 'tested' on the live market, with firm-bespoke licenses, to calibrate regulatory conditions for their final authorization. Performatively and morally, the UK FCA presented these instruments as representative of the fact that it is an experimental regulator (in ways largely consistent with its pre-existing image).

> 'The FCA's regulatory sandbox was a first for regulators worldwide and underlines our deep commitment to innovation and our willingness to think outside the usual regulatory parameters'.[110]

Another consistent aspect of reputation is the performative claim that UK FCA's approaches represent world-leading, unique, and novel solutions for fintechs like cryptocurrency.

> 'We are the first regulator to launch a programme like the sandbox anywhere in the world.... It is an experiment for all involved and we will need to learn as much as the firms engaged in it'.[111]

There were, however, a number of aspects of reputation signalled in cryptocurrency communications which were not present (or not emphasized) in the agency's pre-existing reputation. First, UK FCA more heavily emphasized a moral commitment to facilitating innovation and business development, respectively.[112] Officials overtly characterized Project Innovate as an attempt to make UK FCA more approachable to innovators.[113] Further, UK FCA emphasized its strong performance in developing the sector. Here, UK FCA claims far more direct credit than is seen with NY DFS.

> 'We have seen [sandbox] tests across the full range of sectors that we regulate and I'm pleased that the majority of firms that have tested products in the sandbox have gone on to take the innovation to market'.[114]

99 UK FCA, 'Journey to the FCA.' https://www.fca.org.uk/publication/corporate/fsa-journey-to-the-fca.pdf.
100 UK FCA, 'Business Plan 2-13/14' https://www.fca.org.uk/publication/business-plans/bp-2013-14.pdf.
101 UK FCA (n 99).
102 Eilís Ferran, 'The Break-up of the Financial Services Authority' (2011) 31 Oxford Journal of Legal Studies 455.
103 UK FCA (n 98) 44.
104 R Mashraky, 'FCA Decides Not to Enforce Regulation on Bitcoin | Finance Magnates' (*Finance Magnates | Financial and business news*, 15 December 2017) https://www.financemagnates.com/cryptocurrency/news/fca-decides-not-enforce-regulation-bitcoin/ (last accessed 22 December 2020).
105 Since the period analysed, the FCA has begun to change this stance on cryptocurrencies Rob Davies, 'FCA Proposes Ban on Cryptocurrency Products' (*the Guardian*, 3 July 2019) http://www.theguardian.com/technology/2019/jul/03/fca-proposes-ban-on-cryptocurrency-products (last accessed 22 December 2020).
106 Substantively, cryptocurrencies, wallets, and blockchain applications have been present in multiple rounds of the regulatory sandbox.
107 JA Barefoot, 'Regulation Innovation: The FCA'S Christopher Woolard' 3.
108 'Financial Conduct Authority Unveils Successful Sandbox Firms on the Second Anniversary of Project Innovate' (*FCA*, 7 November 2016) 1 https://www.fca.org.uk/news/press-releases/financial-conduct-authority-unveils-successful-sandbox-firms-second-anniversary (last accessed 23 December 2020).
109 Innovation Hubs are specialized units designed for the purposes of fintech sector engagement and mutual information-sharing.
110 UK FCA, 'Financial Conduct Authority Unveils Successful Sandbox Firms on the Second Anniversary of Project Innovate' (*FCA*, 7 November 2016) 1 https://www.fca.org.uk/news/press-releases/financial-conduct-authority-unveils-successful-sandbox-firms-second-anniversary (last accessed 23 December 2020).
111 Christopher Woolard, 'Innovate Finance Global Summit' (*FCA*, 11 April 2016) 5 https://www.fca.org.uk/news/speeches/innovate-finance-global-summit (last accessed 23 December 2020).
112 This is not to say that FCA was uninterested in criminal activity and consumer protection. Rather, it is a matter of relative emphasis on these aspects in FCA's communications when describing the regulator and its actions.
113 UK FCA, 'Financial Conduct Authority Outlines Lessons Learned in Year One of Its Regulatory Sandbox' (20 October 2017) 1 https://www.fca.org.uk/news/press-releases/financial-conduct-authority-outlines-lessons-learned-year-one-its-regulatory-sandbox (last accessed 23 December 2020).
114 Justin O'Brien, 'Attack on ASIC Chief Draws Corporate Governance into Political Mire' (*The Conversation*, 13 July 2012) http://theconversation.com/attack-on-asic-chief-draws-corporate-governance-into-political-mire-8251 (last accessed 23 December 2020; Greg Medcraft, 'ASIC's Outlook -the Road Ahead' (8 May 2013) https://asic.gov.au/about-asic/

Second, the focus on moral aspects to do with transparency and accountability were not emphasized in this period. Whether this is due to the focus on cryptocurrency communications, or changes over time, is addressed in the discussion.

AUS ASIC

Established in 1998, Australia's Securities and Investments Commission has a longer history of image management than the other regulators. Focusing on the period immediately prior to cryptocurrency, though, we see AUS ASIC presented itself as a procedurally oriented, legalistic regulator (ASIC 2013b). The agency emphasized aspects of appropriate stakeholder consultation and cooperation with other regulators.[115] A focus on procedures, however, ran through all its competencies. AUS ASIC had a performative focus on enforcing financial regulation through litigation; successfully prosecuting a series of high-profile cases. While this might suggest a similar image to NY DFS, AUS ASIC and others characterized its enforcement as 'lawyerly'; cautious and rule-oriented.[116] Another aspect of its performative competencies emphasized was high-quality 'customer-service'. In this regard too, a focus on procedure is apparent, with AUS ASIC issuing charters with detailed standards. In its communications about cryptocurrency, the agency presents a largely similar image.

Like in the UK, cryptocurrencies in the period analysed were not inherently subject to financial regulation.[117] AUS ASIC claimed the regulator had relevant powers where their trade constituted certain kinds of financial goods and services.[118] Despite apparent limits in legal authority, ASIC indicated it had some role in supervising cryptocurrencies. In early 2015, the regulator launched its own Innovation Hub and, in 2016, a regulatory sandbox.[119]

In communications, AUS ASIC presented largely the same procedural, performative, and moral competencies. While AUS ASIC did somewhat reduce its focus on procedural competencies compared with its pre-existing reputation, the agency continued (and far more prominently than in the other two cases) to justify agency decisions by reference to appropriate consultation processes and legal/technical consideration.

> 'In considering the feedback received, we have also consulted with the insurance industry. Based on these discussions, and the submissions received, we consider that the proposed condition is generally workable'.[120]

In discussing cryptocurrencies, ASIC primarily focused on restating its high-quality and ever-improving performance on customer service. The regulator repeatedly discussed improvements to processes, especially in regard to fintech regulatory approvals.

> 'The agreement will enable innovative FinTech companies in Singapore and Australia to establish initial discussions in each other's market and faster and receive advice on required licenses, thus helping to reduce regulatory uncertainty and time to market'.[121]

There are, however, some notable differences in the image ASIC presents in its cryptocurrency communications compared with its pre-existing image. ASIC more heavily emphasizes its performance as a facilitator of business development. Its characterization here is more similar to NY DFS's indirect credit claiming than UK FCA's hands-on involvement.

> 'ASIC supports innovation and we have endeavoured to assist persons to understand their obligations under the laws [regarding digital currency trading] we are responsible for'.[122]

Relatedly, ASIC emphasizes a moral commitment to facilitating innovation not seen in its pre-existing image.

> 'ASIC's fintech licensing exemption reflects our commitment to facilitating innovation in financial services. However, we are equally committed to ensuring that innovative products and services are regulated appropriately and promote good consumer outcomes...'[123]

Another new aspect of its performative reputation is the repeated characterization of its specific approach to the Hub and sandbox was performatively unique and novel.

> 'The proposed licensing exemption compares favourably to measures in other jurisdictions as it will allow some fintech businesses to commence testing of certain product offerings in the absence of detailed assessment by the regulator'.[124]

Also, in regard to uniqueness, in communicating about its performance on cryptocurrency AUS ASIC presented the agency as world-leading in regard to its inter-agency coordination efforts.

> 'Under a new world-first agreement, innovative fintech companies in Australia and the United Kingdom will have more support from financial regulators as they attempt to enter the other's market'.[125]

While this framing reflects a pre-existing reputation for continuously improving procedures, the focus on uniqueness and novelty was not previously strongly emphasized.

news-centre/speeches/asics-outlook-the-road-ahead (last accessed 23 December 2020).

115 O'Brien (n 113); Medcraft (n 113).

116 AUS ASIC, '10-266AD ASIC Releases Stakeholder Survey' (10 December 2010) 11 https://asic.gov.au/about-asic/news-centre/find-a-media-release/2010-releases/10-266ad-asic-releases-stakeholder-survey (last accessed 23 December 2020).

117 David Chau, 'Bitcoin One Step Closer to Being Regulated in Australia' (22 October 2017) https://www.abc.net.au/news/2017-10-23/bitcoin-one-step-closer-to-being-regulated-in-australia/9058582 (last accessed 23 December 2020).

118 Canberra APH, 'Digital Currency—Game Changer or Bit Player' (4 August 2015) 8 https://www.aph.gov.au/Parliamentary_Business/Committees/Senate/Economics/Digital_currency/Report (last accessed 23 December 2020).

119 AUS ASIC, '16-440MR ASIC Releases World-First Licensing Exemption for Fintech Businesses' (15 December 2016) https://asic.gov.au/about-asic/news-centre/find-a-media-release/2016-releases/16-440mr-asic-releases-world-first-licensing-exemption-for-fintech-businesses (last accessed 23 December 2020).

120 AUS ASIC, 'REP 508 Response to Submissions on CP 260 Further Measures to Facilitate Innovation in Financial Services' (Australian Government Australian Securities and Investments Commission 2016) report https://asic.gov.au/regulatory-resources/find-a-document/reports/rep-508-response-to-submissions-on-cp-260-further-measures-to-facilitate-innovation-in-financial-services (last accessed 23 December 2020).

121 AUS ASIC, '16-440MR ASIC Releases World-First Licensing Exemption for Fintech Businesses' (n 118) 1.

122 M Saadat, 'Senate Economics References Committee Inquiry into Digital Currency: Opening Statement' 1.

123 AUS ASIC, '16-440MR ASIC Releases World-First Licensing Exemption for Fintech Businesses' (n 118) 2.

124 AUS ASIC, '16-185MR ASIC Consults on a Regulatory Sandbox Licensing Exemption' (8 June 2016) 1 https://asic.gov.au/about-asic/news-centre/find-a-media-release/2016-releases/16-185mr-asic-consults-on-a-regulatory-sandbox-licensing-exemption (last accessed 23 December 2020).

125 AUS ASIC, '16-194MR Singaporean and Australian Regulators Sign Agreement to Support Innovative Businesses' (16 June 2016) 1 https://asic.gov.au/about-asic/news-centre/find-a-media-release/2016-releases/16-194mr-singaporean-and-australian-regulators-sign-agreement-to-support-innovative-businesses (last accessed 23 December 2020).

Unlike UK FCA, AUS ASIC sought to amend legislation to accommodate the existence of a sandbox. AUS ASIC's sandbox is a sector-wide 'white list' system allowing start-ups only to test new products on temporary licenses.[126] The way AUS ASIC discusses its approach reflects a pre-existing reputational tension between performative responsiveness and procedural correctness. AUS ASIC characterizes its performance as proactive, but only in the sense of identifying matters to be resolved through proper legal procedure.

> 'Your input [on the Innovation Hub] will also help ASIC stay on top of laws that have become impractical or inappropriate as the sector moves forward'.[127]

5.2 Analysis

In all three cases, agencies presented an image in their cryptocurrency communications largely consistent with their pre-existing reputation. In framing their response, there is little evidence regulators sought to drastically rebrand. The image agencies present, however, differs from their pre-existing image in a few, common ways. Agencies signalled new aspects of their image in regard to cryptocurrency/general fintech regulation. All three began to overtly characterize themselves as innovation regulators. To a greater extent than in their pre-existing image, regulators emphasize they are morally committed to, and performing toward, innovation and the development of innovative businesses. Finally, all three emphasize performative uniqueness and novelty in their regulatory approach in cryptocurrency communications. Overall, regulators frame supervision of cryptocurrency as a natural extension of, and bolster to, of their existing regulatory brand.

There are, however, differences between cases. As each agency framed its response in terms of its pre-existing reputation, there were differences in the nature of the image agencies signalled communications on cryptocurrency. NY DFS showed the least change in the image it presented before and after cryptocurrencies. When discussing its new role as a cryptocurrency regulator, further, NY DFS claimed to have exclusive authority over the technology in its jurisdiction, which AUS ASIC and UK FCA did not. Further, UK FCA and AUS ASIC usually discussed cryptocurrencies as part of a broader fintech phenomenon. NY DFS was more likely to refer to cryptocurrency as a stand-alone innovation, although increasingly discusses it as part of 'fintech'.

What may explain why agencies managed their reputation in these ways? To interpret their responses, we draw on the theoretical framework at Figure 1, derived from bureaucratic reputation theory.

One explanation from theory is that regulators respond to innovation, and claim a role in its supervision, when they think they can govern the technology successfully. This is, however, unlikely to be the case for cryptocurrencies. Cryptocurrencies have anonymous users, are generated and traded across borders, and are technically complex and legally ambiguous.[128] It is often unclear, and was certainly in cryptocurrency's early years, whether tokens are currency or financial products and thus, whether financial regulators have jurisdiction.[129]

Regulatory efforts to supervise cryptocurrencies were therefore likely to be difficult, with a high chance of real or perceived failure. That regulators in the case study chose to use highly public communications to claim a role, then, is surprising.

It could be the case that regulators, here, were forced by their political masters into involving themselves in a risky technology. We consider this possible, but unlikely, given each agency in the study has formal, legal autonomy from government. Another explanation is regulators are incompetent at reputation management. They have been insensitive to the risks supervising cryptocurrency posed to their reputation. Our analysis of communications, however, strongly suggests regulators were well aware of the reputational stakes.

> 'However, there are significant, well founded concerns that financial institutions and regulators for that matter are not keeping up with the expectations of consumers for fast, reliable digital transactions. And that's a serious problem that we all need to address with a heightened sense of urgency and focus'.[130]

> 'But I want to reiterate what I said earlier, which is that community expectations have changed. So too have the expectations of the government and the regulator, and even the black letter law. In line with this, we have set out in our Corporate Plan, released last year, our view of 'what good looks like' in the sectors we regulate'.[131]

> 'Innovation can arise from diverse sources, such as start-ups, technology providers as well as regulated firms, including large financial institutions. They all have the potential to challenge existing business models, products and methodologies to benefit consumers and markets as a whole'.[132]

Assuming regulators were sensitive to the considerable risks of supervising cryptocurrencies, this would suggest the risks of silence or inaction on the technology were greater. There is some evidence regulators may have experienced public pressure to act. Cryptocurrencies and their (lack of) supervision was a topic in the media at the time. Anecdotally, much of this coverage was negative; pointing out the risks to consumer protection, systematic stability, money laundering, and the funding of terrorism and the drug trade.[133] In all three jurisdictions, we see examples where politicians, the media, and other audiences call for more regulatory oversight by financial conduct regulators.[134] It would follow that their high-profile communications,

126 AUS ASIC, 'Fintech Regulatory Sandbox' (2018) https://asic.gov.au/for-business/innovation-hub/fintech-regulatory-sandbox (last accessed 23 December 2020).
127 AUS ASIC, '15-211MR Innovation Hub: ASIC Update' (5 August 2015) 1 https://asic.gov.au/about-asic/news-centre/find-a-media-release/2015-releases/15-211mr-innovation-hub-asic-update (last accessed 23 December 2020).
128 Narayan and others (n 19) ix–xxiii.
129 Saadat (n 121).
130 Ben Lawsky, 'Opening Statement. Hearings on the Regulation of Virtual Currency.' (*AVC*, 2014) https://www.youtube.com/watch?v=TZW7R7FPI-JY (last accessed 23 December 2020).
131 AUS ASIC, 'RG 257 Testing Fintech Products and Services without Holding an AFS or Credit Licence (Withdrawn)' (2017) https://asic.gov.au/regulatory-resources/find-a-document/regulatory-guides/rg-257-testing-fintech-products-and-services-without-holding-an-afs-or-credit-licence-withdrawn (last accessed 23 December 2020).
132 UK FCA, 'Financial Conduct Authority Outlines Lessons Learned in Year One of Its Regulatory Sandbox' (n 112).
133 Angela Monaghan, 'Bitcoin Is a Fraud That Will Blow up, Says JP Morgan Boss | Technology | The Guardian' (13 September 2017) https://www.theguardian.com/technology/2017/sep/13/bitcoin-fraud-jp-morgan-cryptocurrency-drug-dealers (last accessed 22 December 2020; Kim Zetter, 'FBI Fears Bitcoin's Popularity with Criminals | WIRED' (9 May 2012) https://www.wired.com/2012/05/fbi-fears-bitcoin (last accessed 22 December 2020).
134 Committee on Banking, Housing, and Urban Affairs, 'Virtual Currencies: The Oversight Role of the U.S. Securities and Exchange Commission and the U.S. Commodity Futures Trading Commission'; David Campbell, 'City Regulator Warns of "Reputational Risks" of Crypto' (*Wealth Manager*, 28 June 2018) http://citywire.co.uk/wealth-manager/city-regulator-warns-of-reputational-risks-of-crypto/a1133481 (last accessed 21 December 2020; Blanca Hartge-Hazelman, 'Glenn Stevens Says Bitcoins Show Promise, but so Did Tulips' (*Australian Financial Review*, 12 De-

and taking on of responsibility, are a rational strategy designed to reassure audiences they were 'on the case' to manage the risks of the technology.[135] The use of a high-profile communications strategy in response to external threats is consistent with findings from Alon-Barkat and Gilad,[136] Moffitt,[137] and Busuioc and Lodge.[138]

To fully understand regulator reputation management in this case, however, one cannot just examine media coverage of cryptocurrencies. One must consider the broader reputational landscape for financial conduct regulators at the time. Cryptocurrencies emerged in the immediate wake of the Global Financial Crisis. The Crisis, it was widely argued, had been triggered by another innovation: over-the-counter derivatives. The invention of this new kind of financial product "shattered the atom of property",[139] with ultimately explosive results. Financial conduct regulators, however, largely failed to detect and understand their seismic implications. Many regulators left the market for these derivatives un- or under- regulated for decades; a major contributor to the Crisis.[140] Most jurisdictions, and certainly those studied, had reformed or were reforming regulatory regimes in this period. This was typically toward stronger, stricter, more prescriptive regulations for financial institutions (e.g. Dodd-Frank in the US, the new Banking Act in the UK, and implementation of Basel III in Australia). Two of the regulators in this study were replacements for predecessors terminated due to their perceived failures (New York Department of Financial Services and the Financial Conduct Authority). AUS ASIC had survived, but still received some criticism for, its handling of the credit market leading up to the Crisis.[141] Financial regulators were at this point, then, on the public mind and likely receiving more scrutiny than in more rosy economic times. It would probably have been far riskier at this moment to try to ignore cryptocurrencies or dodge responsibility.

Regulators may also have chosen high-profile communications strategies, however, in order to shape and manage audience expectations as to the nature of their response.[142] Agencies in our case study do appear to use communications to mitigate the risks of taking on a role in cryptocurrency regulation. There are a number of instances where agencies put boundaries on their obligations and manage expectations about regulatory capacity.

> 'We are regulating financial intermediaries. We are not regulating software development. It's not what we do'.[143]

> 'However, we cannot mitigate every risk, nor do we aim to do so'.[144]

> 'Our response to these developments should be driven by... resisting the temptation to jump before we properly understand developments.'[145]

Indeed, the goal of expectations management may help to explain why all three regulators communicate so little about the technical dimension of reputation. Agencies may seek to moderate expectations about what they could be expected to know about cryptocurrencies, especially in early stages. From this perspective, regulator reputation management is a rational strategy designed to mitigate risks. To respond to media criticism about regulatory negligence, agencies seek to convince their audiences that they are taking swift action to supervise cryptocurrencies. At the same time, they frame responses in ways which temper audience expectations about what can be achieved.

In all three cases, however, in their image management regulators signal not just that they are doing 'something' about cryptocurrency, but that they are doing something extraordinary. The regulators all signal they are unique, novel, and highly successful innovation supervisors. This kind of strategy is irrational if agencies are just managing risks. This kind of public credit-claiming, novelty, and differentiation are high risk communication strategies.[146] They raise expectations. They make agencies a bigger target if anything goes wrong. To help to explain this behaviour, we need to turn to other contextual factors in our framework: agency jurisdictions and pre-existing, unique reputations.

Cryptocurrency trading supervision was relevant to all three financial conduct regulators studied due to risks to – at minimum — consumer protection. None of these regulators, though, necessarily held exclusive jurisdiction over every area of cryptocurrency supervision. NY DFS had a more extensive mandate than UK FCA and AUS ASIC, including powers over criminal investigation, enforcement, and market regulation.[147] In terms of actual instances of jurisdictional competition, in the UK there is little evidence of other agencies trying to claim jurisdiction over UK FCA's traditional regulatory responsibilities (e.g. consumer protection, competition).[148][149] UK FCA actually collaborated with Bank of England and Treasury on a response. For AUS ASIC, we see more competition; notably with other agencies granted formal jurisdiction over certain aspects of cryptocurrency supervi-

cember 2013) https://www.afr.com/policy/economy/glenn-stevens-says-bitcoins-show-promise-but-so-did-tulips-20131213-iygau (last accessed 21 December 2020).

135 see also: Tzur (n 9).

136 'Compensating for Poor Performance with Promotional Symbols: Evidence from a Survey Experiment' (2017) 27 Journal of Public Administration Research and Theory 661.

137 'Promoting Agency Reputation through Public Advice: Advisory Committee Use in the FDA' (2010) 72 The Journal of Politics 880.a

138 Madalina Busuioc and Martin Lodge, 'The Reputational Basis of Public Accountability' (2016) 29 Governance 247, 95.

139 Ford (n 1) 142.

140 Ford (n 1).

141 Hartge-Hazelman (n 133).

142 Gilad, Alon Barkat and Braverman (n 37); Moffitt (n 136) 95.

143 NY DFS, 'Superintendent Lawsky Issues Notice of Intent to Hold Public Hearing Regarding Virtual Currencies on January 28 and 29 in New York City.' (n 90).

144 UK FCA, 'Financial Conduct Authority. Business Plan 2016 / 17 - PDF Free Download' (2017) https://docplayer.net/18378085-Financial-conduct-authority-business-plan-2016-17.html (last accessed 23 December 2020).

145 Greg Medcraft, 'ASIC's Regulatory Approach to High-Frequency Trading and Dark Pool' https://download.asic.gov.au/media/4224331/greg-medcraft-speech-oxford-university-published-24-april-2017.pdf

146 Hood (n 32); David L Deephouse, 'To Be Different, or to Be the Same? It's a Question (and Theory) of Strategic Balance' (1999) 20 Strategic Management Journal 147.

147 In Australia, competition is the responsibility of the Australian Competition and Consumer Commission. In New York it is an obligation of the Antitrust Bureau. Investor protection in the UK and the US is governed by private law, whereas it is public in Australia (and in ASIC's remit). In Australia, money laundering and counter terrorism issues related to currency are the responsibility of the Australian Transaction Reports and Analysis Centre. In the UK, the UK FCA is formally responsible for anti-money laundering but does so as a supervisor of private and professional bodies who engage in the actual enforcement. Counter-terrorism in relation to currency is primarily managed by the Treasury. Both money laundering and counter terrorism matters regarding cryptocurrency are also shared jurisdictions with European Union regulators.

148 Anthony Cuthbertson, 'UK Authorities Lay out What They Will Do about Bitcoin' (*The Independent*, 10 April 2018) https://www.independent.co.uk/life-style/gadgets-and-tech/news/cryptocurrency-bitcoin-regulation-fca-price-updates-market-a8296411.html (last accessed 21 December 2020).

149 UK HM Revenue & Customs did assume responsibility to administer laws about tax and money laundering.

sion.[150] NY DFS experienced jurisdictional incursion from above. The Office of the Comptroller of the Currency discussed offering cryptocurrency companies charters at the federal level, going over heads of state regulators. NY DFS fought this; successfully challenging OCC's charters in court.[151]

That cryptocurrencies were relevant to the core business of financial conduct regulators may help to explain why all three regulators chose a high-profile communication strategy and sought to integrate a role for its supervision into their existing public image. Differences in the nature of NY DFS's jurisdiction and mandate to that of UK FCA and AUS ASIC may also help us to understand how each framed their response. UK FCA and AUS ASIC framed their response in ways that acknowledge the agencies' limited mandate and jurisdiction. They present themselves as having a partial role in the regulation and facilitation of high-tech financial innovation, but lacking legal jurisdiction to singlehandedly regulate cryptocurrencies.[152] NY DFS made a far stronger claim, arguing they were the obvious, exclusive regulator of cryptocurrency trading in its financial conduct aspects. NY DFS may well have communicated as early as it did on cryptocurrencies because of its – obviously founded – fear that other agencies would try to make claims first. It is notable here that NY DFS had more potential competition than AUS ASIC or UFCA. As a state regulator, NY DFS did not only have to guard against encroachments from other agencies in their state but also from federal regulators. Whereas UK FCA and AUS ASIC would likely have had to share authority with other agencies over cryptocurrencies, NY DFS had the potential to supervise largely autonomously. There were, however, other differences in the exact image the three regulators presented; in which dimensions and aspects of reputation they signalled. Bureaucratic reputation theory suggests such differences are likely to arise from differences in their pre-existing reputations.

In our case study, despite the disruptions of cryptocurrency, and its differences to traditional payments, currencies etc., agencies tend to frame their response as an extension of the agency's existing brand. This helps to explain differences in image management between agencies. Why NY DFS presented its responses – certainly initially – as tough, enforcement measures against terrorists and money launders. Why UK FCA presented its response as part of a broader flexible and world-leading strategy on fintech. Why ASIC signalled procedural caution, and a willingness to wait for a new legal mandate to act.

These differences in image management also reflect differences in the unique reputation of each regulator. UK FCA emphasizes that the agency promotes competition through its response to cryptocurrencies, while NY DFS and AUS ASIC do not. Indeed, its role as a competition regulator may help to explain UK FCA's greater focus on innovation and business facilitation in framing its response compared to the other regulators. AUS ASIC repeatedly claims it protects investors, while NY DFS and UK FCA do not directly address investor interests. NY DFS presents itself as a part of the fight against global money laundering and terrorism, a competency to which the other two regulators do not commonly refer. In all cases, these obligations (competition, anti-terrorism, and investor protection) are important parts of each agency's mission statements. These were priorities their governments intended the agencies to address.

In these cases, then, agencies have sought to frame their response to cryptocurrencies to bolster their pre-existing image. In bureaucratic reputation theory, as discussed, this is typically rational behaviour. Agencies have established a reputation which appeals to their audiences prior to innovation and will be reluctant to change a winning formula.[153] In this case, we can make informed speculations about the role of agency audiences in shaping how regulators framed their response to cryptocurrencies. In fact, the composition of audiences for financial conduct regulators helps to explain the new and different aspects of reputation all three agencies do demonstrate.

Finance and banking are sectors dominated by medium-large, highly professionalized institutions (banks, credit unions, corporations etc.) This is what regulators were accustomed to and what regulatory regimes had been designed around. Cryptocurrencies were one of the first fintechs to bring tech start-ups into finance.[154] One might expect this audience has different priorities and preferences for their regulator than large, professional institutional incumbents. The introduction of these new audiences could help to explain why regulators signal they are now innovation supervisors, and why all regulators moved toward a more positive, facilitative tone over time.[155] Regulators may also be trying to frame responses to appeal to existing financial institutions seeking to exploit the opportunities of tech like cryptocurrency.[156] As cryptocurrency proponents become more powerful and influential relative to detractors, one would expect more of the pro-innovation, pro-business framings we do indeed see in this case.[157]

Agency image management, then, could be an attempt to respond to the demands of a burgeoning pro-cryptocurrency coalition. Alternatively, agencies may have been using their communications to construct such a coalition. They framed their response to cryptocurrencies to proactively build support for the agency's preferred course of action, rather in capitulation to audience demands.[158] There are a number of reputational opportunities which may explain such behaviour.

As discussed, novel technologies provide agencies the opportunity to be seen as more unique and valuable to their society. Cryptocurrencies were an opportunity, in particular, for regulators to bolster their reputation in post-Global Financial Crisis period. As discussed, this was a time of reduced trust in traditional financial institutions and their regulators. While this meant that regulators were facing greater scrutiny at this time, it also may have meant they were looking for opportunities to prove themselves. For NY DFS and UK FCA specif-

150 AUSTRAC, 'New Australian Laws to Regulate Cryptocurrency Providers | AUSTRAC' (11 April 2018) https://www.austrac.gov.au/new-australian-laws-regulate-cryptocurrency-providers (last accessed 21 December 2020).

151 Finextra Research, 'New York Defeats OCC in Legal Battle over Bank Charters' (23 October 2019) https://www.finextra.com/newsarticle/34626/new-york-defeats-occ-in-legal-battle-over-bank-charters (last accessed 21 December 2020).

152 Saadat (n 121); Mashraky (n 103).

153 Busuioc and Lodge (n 137).

154 Arner, Barberis and Buckley (n 3) 1305.

155 Maor (n 12); Carpenter (n 11) 33.

156 The payments and money transfer sectors are not monolithic in this regard. One of the most disruptive aspects of cryptocurrencies is their challenge to the hegemonic power of banks and other large financial institutions. Some institutions have responded by demanding regulators ban their competitor. Others sought the freedom to pursue cryptocurrency's commercial applications. Phillip Inman, 'Bank of England to Consider Adopting Cryptocurrency' The Guardian (21 January 2020) https://www.theguardian.com/technology/2020/jan/21/bank-of-england-to-consider-adopting-cryptocurrency (last accessed 22 December 2020).

157 Young (n 60); Rimkut (n 15); Donald P Moynihan, 'Extra-Network Organizational Reputation and Blame Avoidance in Networks: The Hurricane Katrina Example' (2012) 25 Governance 567.

158 Mark C Suchman, 'Managing Legitimacy: Strategic and Institutional Approaches' (1995) 20 Academy of Management Review 571.

ically, cryptocurrencies were an area where they could demonstrate success where their predecessors were seen to have failed. Cryptocurrencies offered an opportunity to demonstrate these agencies could competently manage complex regulatory challenges.

It is notable, further, that regulators tended to frame their responses to cryptocurrency regulation as having a role in *innovation supervision*. Economically, this was a period of high interest and investment in digital technology in general and financial technology in particular.[159] There is evidence that the US, UK, and Australia were all interested in attracting and keeping financial technology in their jurisdiction.[160] Financial technology firms are relatively mobile, not as tethered to geographic locations as businesses with more of a physical presence. Such firms, then, were well placed to engage in regulatory arbitrage.[161] Culturally, technology and 'innovation' have largely positive connotations in those societies (progress, modernity, 'cool'). [162] In societies which value innovation, regulators perpetually stand a lot to gain reputationally from being seen as making a unique, irreplaceable contribution to facilitating the safe and legal trade of novel technologies.[163] The period in which regulators were responding to cryptocurrencies aligns, though, with a renaissance of public interest in – and romanticism of — 'tech' (after the disillusionment of the dotcom bubble bursting in the 1990s).[164] In terms of fintech in particular, the wave of innovation in this period was highly consumer-facing. Unlike previous waves, which mostly affected financial professionals, ordinary people were using and enjoying fintech products. After all, anyone can buy cryptocurrency tokens.[165] The enthusiasm for fintech and public faith in its ability to bring about growth and better quality of life stands in stark contrast to the banal image and lack of public trust in traditional finance. Cryptocurrencies are emblematic of these differences; designed as a decentralized, democratized, reliable, and high-tech replacement for centralized, elite, untrustworthy, unstable, and old-fashioned banking.[166] Public opinion on tech, fintech, and mainstream finance, therefore, may have created a disincentive for regulators to be perceived as opposed to or undermining innovation and growth. Thus, there are historic, economic, cultural, and political reasons that financial conduct regulators might have wanted to realign their public image to include a role in innovation supervision.

This goal would explain why – in our findings — regulators were signalling unique and novel regulatory performance. They were willing to bear the risks of a high-profile failure on cryptocurrencies in order to forge a reputation as an effective innovation supervisor. This goal also explains why all three regulators came to – over time — discuss cryptocurrency more often as part of the broader phenomena of 'fintech' and 'innovation'. Innovation is both a more expansive, and more PR-friendly, framing. Analysing the cryptocurrency case with a bureaucratic reputation framework, then, we see several factors which may explain why regulators chose the reputation management strategies they did. Our findings have implications for both theory and practice.

6. Discussion and conclusion

In this study we examined how regulatory agencies manage their reputation in the face of innovation through a case study of three financial regulators responding to the emergence of cryptocurrency trading. We find all three agencies managed their reputation through a high-profile communications strategy where they discussed their response to cryptocurrency often and in very public fora. In those communications, agencies frame their response as largely consistent with — rather than a radical departure from – their existing public image. Our analysis suggests regulators in this case did not purely see cryptocurrencies as a threat. Rather, they saw opportunities to bolster their reputation in the wake of the Global Financial Crisis.

This paper makes a theoretical contribution by bridging bureaucratic reputation and innovation governance scholarship. We present a theoretical framework to describe and compare how regulators manage their reputation in the face of innovation, and why. Our case study illustrates how — theoretically and methodologically — such a framework can be applied to provide insight into the political motivations and tactics of regulators responding to innovation.[167] Our findings contradict a common assumption that regulators always see innovation in terms of threats.[168] Conversely: that reputational concerns will make regulators reluctant to get involved in the supervision of complex, uncertain new technologies.[169] In the case study, further, we find regulators do not simply react to public demands about technology supervision, but seek to shape those demands. Regulators are independent political actors who use discourse and rhetoric to shape how we see new technologies; their risks, and their opportunities.[170] This demonstrates the value of our theoretical framework over earlier accounts which assume regulators only consider innovation in terms of its risks.[171] Our findings, however, suggest our own theoretical framework should be further expanded. We find that the way regulators responded to cryptocurrency was not just about that technology. It was seemingly about the regulators' broader strategies to build reputation after the damage of the Global Financial Crisis. Thus, in explaining regulator reputation management in response to innovation, we suggest one must also consider the wider political context.

From a practical perspective, regulatory practitioners responding to innovation in their jurisdiction need to be aware of the kind of image they present. When innovative companies see regulators as tough and combative, for instance, this can undermine their willingness to share information and otherwise cooperate with those regulators.[172] Regulatory reputation is a factor which explains why some regulators succeed, and others fail, in their interventions to supervise innovation.[173] From our findings, practitioners should note, in particular, that agencies tend to frame responses as an extension of the regulator's

159 EY (n 79).
160 Philipp Maume, 'Reducing Legal Uncertainty and Regulatory Arbitrage for Robo-Advice' (2019) 16 European Company and Financial Law Review 622; Stijn Claessens and others, 'Fintech Credit Markets Around the World: Size, Drivers and Policy Issues' (Social Science Research Network 2018) SSRN Scholarly Paper ID 3288096 https://papers.ssrn.com/abstract=3288096 (last accessed 22 December 2020).
161 Heikki Marjosola, 'The Problem of Regulatory Arbitrage: A Transaction Cost Economics Perspective' [2019] Regulation & Governance http://onlinelibrary.wiley.com/doi/abs/10.1111/rego.12287 (last accessed 18 February 2021.
162 Ford (n 1) 7–9.
163 Carpenter (n 14).
164 Sara M Smyth, 'The Facebook Conundrum: Is It Time to Usher in a New Era of Regulation for Big Tech?' (2019) 13 International Journal of Cyber Criminology 578.
165 Other kinds of fintech in the current innovation wave – apps, platforms, crowdfunding, roboadvice – are similarly technologies used by ordinary people and not just financial professionals.
166 Davis (n 18).
167 Carpenter (n 11) 754.
168 Maor (n 7); Weaver (n 32); van Erp (n 32); Hood (n 32).
169 Gerding (n 54); Ford (n 1).
170 Carpenter (n 14); Carpenter (n 11); Suchman (n 157); Jones and Millar (n 6).
171 Maor (n 7).
172 Mandel (n 5).
173 Gregory N Mandel, 'Regulating Emerging Technologies' (2009) 1 Law, Innovation and Technology 75; Carpenter (n 11).

existing brand. This may, however, be counter-productive if one's existing brand is at odds with the demands of innovation supervision.

6.1 Limitations and topics for future research

Limitations of the study are, first, its methodological focus on communications about cryptocurrencies rather than all communications published by the agency. While it would have been impractical to qualitatively analyse a decade's worth of agency communications, this allows for the possibility agencies decided to rebrand generally and not just in cryptocurrency communications. Another limitation is that, because Twitter archives tweets, some may not have been available at the time of data collection. Some issues also arose from the coding method. Our method intentionally only captures explicit statements,[174] and not more 'implicit' signalling agencies may have used.[175] This may explain why technical competencies were not commonly signalled: because technical competency is more often 'shown' than it is 'told'. This study collected communications about cryptocurrency in a set period of time, but cryptocurrencies and their regulation are an ongoing and evolving field. Many new developments have emerged since analysis was completed (for example, Her Majesty's Treasury in the UK has launched a consultation on cryptocurrencies in January 2021). The agencies chosen for the case study are not perfectly identical to one another. While we intentionally chose a state over national regulator for the US case to make the cases more comparable in some regards, differences between these two types of regulators could potentially account for differences in NY DFS's choices of reputation management strategy. Finally, responses to radical innovation by three financial regulators may not be representative of all responses by all kinds of agencies in all domains.

Further studies could seek to apply this theoretical framework, and the expectations it implies, to the study of reputation management by other regulators responding to radical innovation in other fields (beyond finance and pharmaceuticals). Theory and research on this topic is still in early stages. More exploratory work is required in a range of regulatory contexts (in-depth case studies, ethnography, discourse analysis etc.). A central question for future research is the extent to which regulatory agencies manage reputation in the face of radical innovation reactively (in response to audience demands) or proactively (attempting to shape audience demands). For the regulators discussed here, a valuable future study would be a media analysis examining of what demands were being made by which stakeholders in these three jurisdictions as a potential explanation for their choice of reputation management strategies. Interview studies with regulator staff could further test the findings of this study, and examine possible reactive and proactive explanations.

174 ASIC, for example, had a pre-existing reputation for procedural correctness. Its communications used far more distant, technical language; more commonly entered around questions of law. This implicit signalling of procedural competency could not be captured in this study.

175 e.g. Kjersti Thorbjørnsrud, 'Mediatization of Public Bureaucracies: Administrative versus Political Loyalty' (2015) 38 Scandinavian Political Studies 179.

APPENDIX A Detailed methodology and results of coding

In this study we compare reputation management responses of three financial regulators (NY DFS, UK FCA, and AUS ASIC). We examined which communication strategy each agency chose and whether, and how, they engaged in image management. Image management was determined through comparing the image they presented in their communications about cryptocurrency to their image in the period immediately prior, then comparing between cases.

The study used three methods: 1) qualitative document review of the agency's pre-existing image and 2) quantitative and 3) qualitative content analysis of cryptocurrency communications. The quantitative analysis determined communications strategy. The document analysis, with the qualitative content analysis, analysed image management.

For the document analysis, we searched Google Scholar, Westlaw, and Lexis Nexis with agency titles, acronyms, and 'reputation'. Documents were included if they were published in the three years prior to the agency's first communication about cryptocurrency. Documents included the agency's own statements, academic literature, and authoritative media and expert judgements. To determine the nature of the agency's pre-existing image, documents were interpreted using the coding schema described below.

For the quantitative content analysis, we collected all agency communications published after 2008 and before March 2018 about cryptocurrency or closely related topics like general statements about fintech (where cryptocurrency was a technology under that label). We searched agency websites and official Twitter account(s)[176] with the word cryptocurrency and closely associated terms. We collected 538 individual texts. These were imported into NVIVO and analysed to determine text type (e.g. speech, tweet) and audience (e.g. mass, private).[177] Agencies were considered to have chosen low- or high- profile strategy based on number of texts, frequency of publishing, and high- versus low- profile fora (e.g. targeted, private speeches versus media appearances).

A stratified (by type) random sample of 351 texts were then subjected to qualitative content analysis to determine what kind of image each agency presented. We developed a coding schema using Carpenter's framework of reputational competencies and informed by previous analyses using that framework.[178] This is summarized in Table A1. After coding we conducted a summative analysis of the documents. We determined roughly which kinds of competencies and aspects agencies raised most often. These aspects were then interpreted qualitatively to determine the overall image the agency was constructing.[179] This was then compared with the competencies and aspects presented by the other two agencies, and compared to its pre-existing image. Summary results by agency are presented in Tables 2-4.

176 @DFS, @TheFCA, @ASICMedia, @ASIC_Connect, @MoneySmartTeam
177 Moschella and Pinto (n 66) 520.
178 e.g. Rimkut (n 15).
179 Hsieh and Shannon (n 82) 124–5.

Table A1 Coding schema

	Description	Agency examples	'Action' examples	' Goal' examples
Performative	Phrase refers to capacity of the agency to achieve desired outputs and outcomes; the extent to which it is substantively successful – including efficiency.	We are an effective and efficient market regulator.	Improvements to the regulatory framework has attracted foreign investment. By updating our procedures, we have reduced financial licensing fees by 10%.	Increasing market competition is our central goal. We will publish regulatory guidance in the next quarter.
Technical	Phrase refers to the expertise of the agency relevant to its capacity to perform its role; examples: "scientific accuracy, methodological prowess, and analytical capacity".	The staff of our innovation unit are experts in fintech. The agency is still learning about fintech.	The current policy is based on a quantitative analysis of market trends in 8 jurisdictions. We are implementing a sandbox to gather evidence about regulatory effectiveness.	The agency aims to increase its analytical capacity by establishing a specialist 'market scanning' unit.
Procedural	Phrase refers to the use of correct procedures associated with decision making: • Procedural fairness • Adequate evidence collection and provision • Decisions based on evidence • Meeting consultation requirements • The thoroughness of procedures.	The agency acts in accordance with the requirements of the Administrative Proceedings Act 1959.	Our enforcement decision against [company X] was made in accordance with Guidelines v3.1.	The agency will increase consultation periods from 2 to 4 weeks.
Moral	Phrase refers to the ethics or morality of the agency's goals or means, including: • Protecting the interests of stakeholders • Honesty • Kindness • Compassion • 'Humanity'.	We consider ourselves a guardian of competitive markets. The agency considers itself a partner to industry, helping firms to comply.	We have published the risk analytics to enable transparent debate about the risks of [policy X].	We are committed to maintaining an even playing field for all firms. Our goal is to protect consumers.

Table A2. Image signalled by NY DFS in cryptocurrency communications

	Aspects From Pre-Existing Image	Additional Aspects
Performative	Is tough, stringent, and comprehensive in market supervision; gets results Is more effective than federal regulators	Performs well in regulating cryptocurrency/financial innovation Implements unique and novel regulatory solutions Regulation not hindering (indirectly helps) facilitate business development Regulation not hindering (indirectly helps) facilitate financial innovation
Moral	Primarily aims to protect consumers of financial products from fraud and other harm Aims to combat illegal activity in New York, the US, and internationally (money laundering and terrorism) Promotes fairness in financial markets; setting appropriate and consistent regulatory standards	Aims to protect consumers/combat illegal activity in regard to cryptocurrency Aims to facilitate financial innovation
Procedural		Makes decisions based on rigorous fact finding and inquiry
Technical	[Not emphasized, rarely discussed]	[Not emphasized, rarely discussed]

Table A3. Image signalled by UK FCA in cryptocurrency communications

	Aspects From Pre-Existing Image	Additional Aspects
Performative	Employs principles/outcomes-based regulation; flexible and adaptable Regulates in ways which promote competition in financial markets, but also protect consumers Supervises proactively; addressing new regulatory issues early Leads the world in creative regulatory solutions	Directly facilitates business development Performs well in regulating cryptocurrency/financial innovation Regulator directly facilitates financial innovation
Moral	Has a role in promoting market integrity and consumer protection Has a central role in promoting competition, which is balanced with protecting consumers	Aims to facilitate financial innovation
Procedural	Not rigidly rule bound Coordinates their actions with other regulators/agencies	
Technical	[Not emphasized, rarely discussed]	[Not emphasized, rarely discussed]

Table A4. Image signalled by AUS ASIC in cryptocurrency communications

	Aspects From Pre-Existing Image	Additional Aspects
Performative	Supervises proactively, addressing new regulatory issues early through legal procedures Provides high quality 'customer' service to individuals and businesses it regulates or advises	Performs well in regulating cryptocurrency/financial innovation Regulator indirectly facilitates business development Regulator indirectly facilitates innovation Implements unique and novel regulatory solutions Leads the world in inter-regulator coordination on fintech
Moral	Aims to promote the interests of shareholders/other investors Aims to promote fairness in financial markets; setting appropriate and consistent regulatory standards	Aims to facilitate innovation
Procedural	Coordinates appropriately with other regulators Facilitates stakeholder deliberation where issues not resolved in law	
Technical	[Not emphasized, rarely discussed]	[Not emphasized, rarely discussed]

02

Technology and Regulation

Not Hardcoding but Softcoding Data Protection

Aurelia Tamò-Larrieux*, Simon Mayer** & Zaira Zihlmann***

GDPR, privacy by design, techno-regulation, privacy engineering, softcode

aurelia.tamo@unisg.ch

simon.mayer@unisg.ch

zaira.zihlmann@unilu.ch

The delegation of decisions to machines has revived the debate on whether and how technology should and can embed fundamental legal values. In this article, we discuss the translational, system-related, and moral issues raised by implementing legal principles in software. While our findings focus on data protection law, they apply to the interlinking of code and law across legal domains. These issues point towards the need to rethink our current approach to design-oriented regulation and to prefer 'soft' implementations, where decision parameters are decoupled from program code and can be inspected and modified by users, over the 'hard' embedding of such parameters into opaque pieces of program code.

1. Introduction

With more smart devices guiding us through our daily activities comes the quest to ensure that these technologies reflect the fundamental values of the society they are embedded in. Smart products like social robots can sense their environment, weigh various options against each other, and act upon their decision-making.[1] The key question thus becomes how options within the decision-making process are balanced and whether those decisions can take the legal environment into account.

The automatic adaptation of code to the legal parameters set out in law raises fundamental questions. A rich literature on techno-regulation and hardcoding or hardwiring data privacy exists, upon which this article builds.[2] Whether the encoding of law appears as part of the 'solution space'[3] or part of a problem, depends also on what legal field one is analyzing (e.g., Intellectual Property rights and Digital Rights Management systems, privacy by design).[4] What is clear is that law in writing vs. law in code can have very different properties, i.e., act differently upon society, thereby raising *systemic* and *moral* issues.

While interdisciplinary research groups have been active in addressing *translational* challenges of interlinking code and law,[5] philosophers and legal scholars have debated the merits and limitations of such initiatives. Seminal research has been conducted among others by Ronald Leenes, who has disentangled techno-regulatory initiatives originating from state and non-state regulators;[6] Mireille Hildebrandt, who has coined the term 'Ambient Law' which more broadly strives to integrate legal protection into the design of technology;[7] Karen Yeung, who analyzes the different effects of legal prohibition vs. techno-regulation on moral agency suggesting that the partial erosion of moral

1 George A. Bekey, 'Current Trends in Robotics: Technology and Ethics' in Patrick Lin, Keith Abney and George A. Bekey (eds), *Robot Ethics: The Ethical and Social Implications of Robotics* (MIT Press 2012) 17.

2 Lee Bygrave, 'Hardwiring Privacy' in Roger Brownsword, Eloise Scotford and Karen Yeung (eds), *The Oxford Handbook of Law, Regulation and Technology* (OUP 2017) 754; Bert-Jaap Koops and Ronald Leenes, 'Privacy regulation cannot be hardcoded. A critical comment on the 'privacy by design' provision in data-protection law' (2014) 28(2) *International Review of Law, Computers & Technology* 159; Ugo Pagallo, 'On the Principle of Privacy by Design and Its Limits: Technology, Ethics and the Rule of Law' in Serge Gutwirth and others (eds), *European Data Protection: In Good Health?* (Springer 2012) 343; Karen Yeung, 'Can We Employ Design-Based Regulation While Avoiding Brave New World?' (2011) 3(1) *Law, Innovation and Technology* 1.

* Aurelia Tamò-Larrieux is an International Postdoctoral Fellow at the Law School of the University of St.Gallen.

** Simon Mayer is a Professor of Interaction- and Communication-based Systems at the Institute of Computer Science of the University of St.Gallen.

*** Zaira Zihlmann is a PhD Candidate at the Faculty of Law of the University of Lucerne.

Received 13 Aug 2020, Accepted 10 Apr 2021, Published: 15 May 2021.

3 Urs Gasser, 'Recoding Privacy Law: Reflections on the Future Relationship Among Law, Technology, and Privacy' (2016) 130(2) *Harvard Law Review Forum – Law, Privacy & Technology Commentary Series.*

4 Bygrave, 'Hardwiring Privacy' (n 2), 755.

5 Cf. e.g., Ronald Leenes and others, 'ENDORSE. Deliverable D2.5 Legal Requirements' (2011) https://cordis.europa.eu/docs/projects/cnect/3/257063/080/deliverables/001-ENDORSED25submitted.pdf (accessed 29 October 2020); Stefan Schiffner and others, 'Towards a Roadmap for Privacy Technologies and the General Data Protection Regulation: A transatlantic initiative' in *Proceedings of the Annual Privacy Forum 2018* (Barcelona, Spain, June 2018) https://people.cs.kuleuven.be/~bettina.berendt/Papers/schiffner_et_al_APF_2018.pdf (accessed 8 November 2020); Michael Birnhack, Eran Toch and Irit Hadar, 'Privacy mindset, technological mindset' (2014) 55(1) *Jurimetrics* 55.

6 Ronald Leenes, 'Framing Techno-Regulation: An Exploration of State and Non-State Regulation by Technology' (2011) 5(2) *Legisprudence* 143.

7 Mireille Hildebrandt, 'A Vision of Ambient Law' in Roger Brownsword and Karen Yeung (eds), *Regulating technologies: Legal futures, regulatory frames and technological fixes* (Hart Publishing 2008) 175; Mireille Hildebrandt, 'Legal Protection by Design: Objections and Refutations' (2011) 5(2) *Legisprudence* 223.

freedom through technology does not have to result in overall collapse of moral foundations;[8] as well as Emre Bayamlıoglu and Ronald Leenes, who describe how data-driven decision-making that enacts regulatory orders undermines the rule of law.[9]

Guided by a concrete implementation of data protection principles in a smart product[10] and building upon literature on the failures of hardcoding privacy[11], we explore the pitfalls of bottom-up implementations of legal principles into software. This leads to a better understanding of why encoding data protection is an imperfect remedy. Sometimes, the imperfectness originates from the structure and behavior of law, sometimes from the structure and behavior of code. Our goal is to enable a differentiated discussion on those interactions in the specific field of data protection. The translational issues raised throughout the article lead to a call for action for both, the computer science and the legal community. Beyond these translational issues, we discuss systemic and moral challenges raised by design-based regulation. These challenges point to more fundamental questions on how and when we want law to be interlinked with code in a way that code regulates human and machine transactions. We argue that, to address those latter issues, we need to move towards 'softcoding' which decouples decision parameters (e.g., production rules, conditionals, thresholds) from opaque program code and thereby allows users to observe and adapt them. Softcoding does not only lead to advantages on the technology side, since it ensures that systems remain flexible to changes of the (legal) environment; it also entails that systems remain transparent, contestable, and malleable and thereby still allow for disobedience as well as control by users and judges.

This article contains three main sections. In Section 2, we start by describing the design implications of the GDPR with focus on the norm on data protection by design and default. From this overarching principle we move towards discussing hard and softcoding approaches to law as well as the technology implementations that have been proposed to comply with the principles of data protection law. This literature review situates the topic of this article into both its legal and technology contexts. Moving away from this dichotomy, Section 3 discusses why encoding data protection principles in practice is an imperfect remedy. On a meta-level, the imperfectness is grouped into *eight clusters of issues* that arise when taking a bottom-up approach to encoding data protection. Within each cluster, detailed specifications on why the interlinking of code and law does not lead to an isomorphic representation of the foundation of the law within code are discussed. Upon this basis, Section 4 describes a path forward: While in our opinion imperfectness does not equal failure nor suggests that we should abandon those approaches altogether, we emphasize the need for *more flexible, loosely coupled*, implementation approaches that allow for more transparency, contestability, and malleability. We furthermore emphasize the need for transdisciplinary experts who promote responsible technology that does not merely lead to superficial implementations of law in code but to one that preserves core tenets of our legal system. If, in the future, law becomes even more computable[12], then the need to establish clear procedural rules on how to contest hard- or softcoded provisions, ensure understandability of legally binding decisions will become key. Such challenges can only be addressed when moving beyond strictly disciplinary approaches.

2. From an Ideal to Implementations

2.1 "Yes, but..." and Other Design Implications of the GDPR

The quest to interlink law and code and create computable laws is seen in various legal fields such as in data protection law, which will be the focus of this article. As a regulation, the GDPR can be best described as a *compromise*. It is a compromise between different data protection regimes within the EU as well as a compromise between various interests that have shaped its final scope.[13] The compromise between different data protection regimes in the EU was already apparent within Directive 95/46/EC[14] (Directive), the predecessor of the GDPR. The Directive itself drew heavily from the Convention for the Protection of Individuals with regard to Automatic Processing of Personal Data (Convention 108)[15], which was originally signed in 1981 and later updated in 2018. Convention 108 was the initial push to a harmonized data protection approach in the EU.[16] Its main principles were incorporated and refined in the Directive and adopted within the GDPR. Convention 108, the Directive, and the GDPR all outline their 'objectives' and 'purpose' along the lines of wanting to ensure the protection of fundamental rights and freedoms of individuals with respect to their 'right to the protection of personal data'[17] and 'right to privacy.'[18] The objective of protecting fundamental rights is also what makes the application and, as will be shown, technical implementation of data protection law challenging. Fundamental rights in their core *require a balancing approach*, which from a technical perspective means that more often than not the solution will be not merely 'yes' or 'no' but a 'yes, but' or 'no, but' (i.e., its logic is defeasible). The *'yes, but'-principle* is inherent to the European data protection approach.[19]

The principles set in place within Article 5 of the GDPR set the basic

8 Yeung (n 2), 27.
9 Emre Bayamlıoglu and Ronald Leenes, 'The 'rule of law' implications of data-driven decision-making: a techno-regulatory perspective' (2018) 10(2) *Law, Innovation and Technology* 303 et seqq.
10 Kimberly Garcia and others, 'Towards Privacy-Friendly Smart Products' (2021) preprint available here https://www.alexandria.unisg.ch/262898/1/TechPaperToyRobot_Alexandria.pdf (accessed 5 April 2021). See Section 2.2 "Hard- or Softcoding Law" for further context.
11 Koops and Leenes (n 2), 159; Ronald Leenes and Federica Lucivero, 'Laws on Robots, Laws by Robots, Laws in Robots: Regulating Robot Behaviour by Design' (2014) 6 *Law, Innovation and Technology* 193.
12 We understand the term "computable" as used in social science literature as regulation processed by and through machines, while not referring to the theory of computation in computer science.
13 Cf. Ece Ö Atikcan and Adam W Chalmers, 'Choosing lobbying sides: the General Data Protection Regulation of the European Union' (2019) 39(4) *J Pub Pol* 543, 545; cf. also Jukka Ruohonen, 'David and Goliath: Privacy Lobbying in the European Union' (2019) https://arxiv.org/pdf/1906.01883 (accessed 28 October 2020).
14 Directive 95/46/EC of the European Parliament and of the Council of 24 October 1995 on the protection of individuals with regard to the processing of personal data and on the free movement of such data, OJ 1995 L 281/31.
15 Convention for the Protection of Individuals with regard to Automatic Processing of Personal Data 1981, ETS 108.
16 Eleni Kosta, *Consent in European Data Protection Law* (Nijhoff Studies in European Union Law, BRILL Martinus Nijhoff Publischers 2013) 24 with reference to Frederick W Hondius, *Emerging data protection in Europe* (Elsevier 1975) 63 et seqq.
17 Art. 1(2) GDPR and Rec. 1 referring to Art. 8(1) of the Charter for Fundamental Rights of the European Union (Charter); note that the term 'privacy' is not used any longer within the GDPR unlike its predecessor and Convention 108.
18 Art. 1(1) Directive 95/46/EC; Art. 1 Convention 108.
19 Serge Gutwirth and Paul De Hert, 'Regulating Profiling in a Democratic Constitutional State' in Mireille Hildebrandt and Serge Gutwirth (eds), *Profiling the European Citizen: Cross-Disciplinary Perspectives* (Springer Netherlands 2008) 279: "As a rule, personal data may be processed, provided the data controller meets a number of conditions. The rule is a 'yes, but ...' rule.".

rules for processing personal and sensitive data. They contain technical requirements, such as ensuring the integrity and confidentiality of data, as well as ones that demand a balance between the input and output, such as limiting the data collection to what is necessary to achieve a specified purpose. Any data controller must comply with the principles and demonstrate compliance with the principles.[20] The requirement of demonstrating compliance shows that there is no 'right or wrong' implementation of the principles but that their implementation must depend on the *specific case* and the involved risks.[21] In other words, because of the context-specificity multiple ways to implement the data protection principles can co-exist, with some more right or wrong where a definitive answer can only be provided when taking the circumstances, purposes, risks, and remedies into account. Article 5(2) of the GDPR also highlights the personal responsibility of the data controller to determine the adequate measures for the intended data processing.[22] Thereby, Article 5(2) 'serves as a *meta-principle*' as it does not only establish a substantive responsibility of complying with the fundamental principles but also entails a *procedural requirement* of being able to demonstrate such compliance.[23]

The principles are coupled to the *requirement of legality*.[24] The requirement of legality mandates a lawful basis for the processing of personal or sensitive data. The interplay between principles and the requirement of legality found within the GDPR are the product of the compromised approach to data and privacy protection in Europe. As the evolution of data protection law among European countries shows, the approaches in different countries (and later member states adopting the Directive) varied,[25] and to this day influence the interpretation of national courts.[26] From a design perspective such a heterogeneous landscape and understanding of data protection law has engineering implications: Either one designs a system to comply with the (internationally) highest standard of the legal requirements or product variants are built that can adapt to the local regulatory environments.

With the GDPR the focus shifted more and more towards implementing data protection through organizational and in particular technical measures.[27] The implementation of Article 25 of the GDPR introduced the concept of *data protection by design*[28] *and default* into data protection law and thereby requested data controllers to employ technical and organizational measures not only to protect personal data from attacks, leaks, or destruction but overall to ensure that the data protection principles are adhered to. Data controllers must ensure that their engineers and developers implement adequate solutions to protect personal data into their products and services.[29] Failures to include proper measures can result in high fines, as seen in Germany where a company failed to ensure the erasure of personal data of employees (e.g., salary statements, contracts, etc.).[30] Yet, the implementation of technical and organizational measures has its boundaries: The implementation must economically and technically be feasible and the relationship between the risk of the processing and the data protection by design measures set in place must be balanced. In other words, data controllers are not required to "spend a disproportionate amount of resources when alternative, less resource

20 Art. 5(2) GDPR.

21 Horst Heberlein, 'Art. 5 Grundsätze für die Verarbeitung personenbezogener Daten' in Eugen Ehmann and Martin Selmayr (eds), *DS-GVO: Datenschutz-Grundverordnung: Kommentar* (2nd edn, Beck'sche Kurz-Kommentare, C.H. Beck, LexisNexis 2018) 29; European Data Protection Supervisor, 'A Preliminary Opinion on data protection and scientific research' (6 January 2020) https://edps.europa.eu/sites/edp/files/publication/20-01-06_opinion_research_en.pdf (accessed 28 October 2020); Peter Hustinx, 'EU Data Protection Law: The Review of Directive 95/46/EC and the General Data Protection Regulation' in Marise Cremona (ed.), *New technologies and EU law* (The collected courses of the Academy of European Law, Oxford University Press 2017) 154; Milda Macenaite, 'The "Riskification" of European Data Protection Law through a two-fold Shift' (2017) 8(3) *European Journal of Risk Regulation* 506, 525.

22 Heberlein (n 21), 29; Art. 5(2) GDPR refers to "accountability" in the English version of the GDPR, the German wording is "Rechenschaftspflicht" and French wording "résponsabilité"; Lachlan Urquhart and Jiahong Chen, 'On the Principle of Accountability: Challenges for Smart Homes & Cybersecurity' (2020) https://arxiv.org/pdf/2006.11043 (accessed 28 October 2020) 3 et seqq.

23 Urquhart and Chen (n 22), 3 et seqq.; note that Lachlan Urquhart, Tom Lodge and Andy Crabtree, 'Demonstrably doing accountability in the Internet of Things' (2019) 27(1) *International Journal of Law and Information Technology* 1, 10 argue that Art. 5(2) GDPR must be read in conjunction with Art 24 GDPR thereby extending the requirement of (demonstrating) compliance to the whole GDPR.

24 Note that in the EU the principle of lawfulness (Art. 5(1)(a) GDPR) can be interpreted broadly or narrowly. If interpreted narrowly, fulfilling the principle of lawfulness requires establishing an adequate legal ground listed in Art. 6 GDPR. If understood broadly, lawfulness means that no other legal obligations related to the processing of data may be breached and that aside from its legal grounds according to Art. 6 GDPR must be demonstrated. Cf. on said discussion Eike Michael Frenzel, 'Art. 5 Grundsätze für die Verarbeitung personenbezogener Daten' in Boris P Paal and Daniel A Pauly (eds), *Datenschutz-Grundverordnung, Bundesdatenschutzgesetz* (Beck'sche Kompakt-Kommentare, 2nd ed. C.H.Beck 2018) 14 et seqq.

25 Viktor Mayer-Schönberger, 'Generational development of data protection in Europe' in Philip Agre and Marc Rotenberg (eds), *Technology and privacy: The new landscape* (MIT Press 1997).

26 Cf. Rebecca Wong, 'The Data Protection Directive 95/46/EC: Idealisms and realisms' (2012) 26(2-3) *International Review of Law, Computers & Technology* 229, 230; cf. Orla Lynskey, 'The 'Europeanisation' of Data Protection Law' (2017) 19 *Cambridge Yearbook of European Legal Studies* 252, 264 et seqq.

27 While the Directive 95/46/EC already obliged controllers to "implement appropriate technical and organizational measures to protect personal data" (Art. 17 Directive 95/46/EC) its focus rested predominantly on security measures. Nonetheless, courts such as the European Court of Justice (ECJ) already had indirectly required privacy-friendly modifications, such as in the Google vs. Spain decision (C-131/12) which required Google to enable de-indexation (which can be seen as a more privacy-friendly operation). Lee Bygrave, 'Article 25. Data protection by design and by default', *The EU general data protection regulation (GDPR): A commentary* (OUP 2020) 575.

28 The idea of data protection by design aligns with Article 8 of the Charter of Fundamental Rights of the EU which requires the adoption of "technical and organizational measures" to ensure "effective protection." The European Court of Human Rights (ECtHR) also embraced privacy by design ideals in its I v Finland decision. Bygrave, 'Article 25. Data protection by design and by default' (n 27), 575 and I v Finland App no 20511/03 (ECtHR, 17 July 2008) rec. 41 et seq.; Axel M. Arnbak, *Securing private communications: Protecting private communications security in EU law: fundamental rights, functional value chains and market incentives* (Doctoral Thesis, University of Amsterdam IViR 2015).

29 Mireille Hildebrandt and Laura Tielemans, 'Data Protection by Design and Technology Neutral Law' (2013) 29(5) *Computer Law & Security Review* 509, 517; cf. also Lee Bygrave, 'Data Protection by Design and by Default: Deciphering the EU's Legislative Requirements' (2017) 1(02) *Oslo Law Review* 105, 114; Fabian Niemann and Philipp Scholz, 'Privacy by Design and Privacy by Default - Wege zu einem funktionierenden Datenschutz in Sozialen Netzwerken' in Falk Peters, Heinrich Kersten and Klaus-Dieter Wolfenstetter (eds), *Innovativer Datenschutz* (Duncker & Humblot 2012) 109 et seqq.

30 Berliner Beauftragte für Datenschutz und Informationsfreiheit, 'Berliner Datenschutzbeauftragte verhängt Bussgeld gegen Immobiliengesellschaft' (5 November 2019) https://www.datenschutz-berlin.de/fileadmin/user_upload/pdf/pressemitteilungen/2019/20191105-PM-Bussgeld_DW.pdf (accessed 28 October 2020). Smaller fines have been issued based on Art. 25 GDPR in Bulgaria, Greece, Romania. For further cases see GDPR Enforcement Tracker https://www.enforcementtracker.com/# (accessed 28 October 2020).

demanding, yet effective measures exist."[31]

While the scope of Article 25 of the GDPR includes all the principles of the GDPR (i.e., meeting all the requirements of the law) and can thus be seen as a 'hollow norm,'[32] the data protection by design norm differentiates among factors that support 'extra' technical measures or that tip the balance in favor of the data subject and factors that reduce the need to implement technical measures. The former (i.e., factors supporting extra measures) include: *high risks for or impact on the data subject's rights and freedoms*, *'unreasonable' purposes*, and *sensitive context of the processing*. The latter (i.e., factors reducing the burden of implementing technical and organizational measures) include: *costs of the actual implementation of the technical measures and limited scope of the processing* (tied to the purposes of the processing and legitimacy of the purposes). While not strictly mandated by the GDPR, ways to ensure that devices comply with the principles via their software have been promoted by developers (see Section 2.3 "Machine-understandable Data Protection Law"). These approaches encode the principles into devices and try to determine ways to automatically factor in the heterogeneous requirements demands mentioned; however, this requires the creation of complex technical systems.

2.2 Hard- or Softcoding Law

In the aim of a bottom-up approach this article draws on a case study in which a toy robot prototype was developed as a (fictional) learning tool for young children.[33] By taking a toy robot as a use case, one can examine how the legal environment of such a smart product is reflected in its firmware implementation. A toy robot, as will further be elaborated below (see Section 3 "Encoding Data Protection Law: An Imperfect Remedy"), includes various data processing capabilities that challenge the fundamental principles of data protection law (e.g., privacy-sensitive sensors such as cameras, continuous processing of personal data, movable, and used by vulnerable users such as children in their private homes). The design of a toy robot prototype requires an iterative approach, starting from the technical dimensions, considering the data-protection-relevant data flows of the toy robot, and establishing a continuous feedback loop between legal scholars and computer scientists to adapt and augment the data flows of the toy robot to fit the requirements laid out by the law. Those attempts target not only the configuration of the robot itself, but also impact the decision criteria that the robot relies on and on a run-time level the adaptability of the toy robot to changed circumstances.

Privacy-by-design scholars and computer scientists working on machine-understandable data protection law seem to agree that a successful encoding of data protection principles for a given system requires (1) a general description of foundational legal principles, (2) the ability to collect information about legally relevant criteria at run time, (3) specific context- and capacity-tailored decision criteria of how the principles (1) are applied together with the criteria (2), and (4) the ability to act upon the decisions produced by (1-3) by adapting the system's behavior at run time. When designing a privacy-friendly toy robot, (1) is satisfied by building upon available ontologies[34] (see below Section 2.3 "Machine-understandable Data Protection Law"; e.g., the concept of parental consent). (2) is given when the robot obtains context data through its virtual or physical sensors (e.g., the data subject's age or the robot's current location). (3) evaluates the legal principles (from (1)) given the context data (from (2)); e.g., to determine Member State specific parental consent age limits, or information about the data subject's age). And (4) is established when the robot is able to update its procedures when circumstances change (e.g., when the robot moves to a new jurisdiction, or parental consent is not required anymore).

To better distinguish between the different components and implementations of data protection by design approaches we start by the norm addressee: While Article 25 of the GDPR binds data controllers, the implementation in particular of technical measures will rest upon the engineers and developers creating the data processing devices.[35] If developers want to configure a product that adheres to the fundamental principles of data protection law, many design decisions will have to be taken already at the time of designing the software architecture of the system and implementing its software modules and they will need to consider the advice of legal experts. For instance, determining the possible legal grounds for processing, the purposes of processing, or the possible ways and technical means to adhere to the principle of transparency, the minimization of data and limitation of storage, as well as the implementation of security principles will have to be determined when developing a smart product and implemented into the design from the beginning. However, developers can choose to design a robot that does not only reflect a single set of pre-defined purposes or legal contexts but can select among (not: decide or judge) at run time which among a multitude of different possible settings it adopts. In other words, data controllers define collections of parameters with legal implications together with heuristics that allow the robot to select one of these - in this way, the robot can - at run time - adapt to legal, contextual, and technical changes. For instance, a legal change would occur if a smart device moves from one jurisdiction to the other and the age of consent changes (e.g., from France, where the consent age is 15, to Belgium where the consent age is 13). Adapting to this change would require access to the geolocation of the device (component (2) above) in order to ensure that the robot requires a new consent (components (3) and (4)) if the age threshold has changed according to the shared understanding of legal principles (component (1) above). Or as another example, if new security standards are published a robot could automatically change its processing operations to adhere to these new standards (e.g., encryption standards) - this is referred to as "crypto-agility"[36] but follows a very similar architecture in that shared foundational assumptions need to be laid out in a machine-readable way and used as a basis for the contextual adaptation of the system's behavior at run time. Thus we see that this configuration impacts the behavior of the toy robot at run time. This does, however, not make the toy robot per se a norm addressee of the GDPR but merely is a way for data controllers, via their engineers, to ensure that their devices are tailored to local requirements in data protection law and can adapt

31 EDPB, 'Guidelines 4/2019 on Article 25 Data Protection by Design and by Default Version 2.0' (20 October 2020), at 9.

32 Aurelia Tamò-Larrieux, *Designing for Privacy and its Legal Framework: Data Protection by Design and Default for the Internet of Things* (Issues in Privacy and Data Protection, Springer International Publishing 2018) 209.

33 Kimberly Garcia and others (n 10). The toy robot roams private family rooms, taking pictures of its surroundings every few seconds and analyzing them to identify known people (typically children) within its field of view. Once a person has been identified, the robot stops to perform an educational action, such as playing a song that motivates the identified person to sing along using preselected personalized content, which would be tailored to the child's age and current interests.

34 E.g., DPV3 vocabulary, https://dpvcg.github.io/dpv/ (accessed 28 October 2020).

35 Tamò-Larrieux (n 32), 84 et seq.

36 Bryan Sullivan, 'Cryptographic Agility' (2010) available at http://media.blackhat.com/bh-us-10/whitepapers/Sullivan/BlackHat-USA-2010-Sullivan-Cryptographic-Agility-wp.pdf (accessed 20 December 2020).

over time to new requirements automatically.

This is where the distinction between hardcoding and softcoding comes in. Above, we introduced "softcoding" as the decoupling of decision parameters (e.g., production rules, conditionals, thresholds, etc.) from opaque program code. We argue that this would better enable users to understand, monitor, and adapt systems compared to the "hardcoded" implementation of regulation directly in program code. The inflexibility that this entails does not only have negative consequences regarding the *adaptivity of a system*: Assuming that a device has hardcoded rules, updating the device to for instance a changed legal landscape (e.g., from German to Swiss data protection law) would require sending in the product to upload a different variant of the software. Via softcode, these rules could instead be retrieved at run time and could even be kept up to date with current decisions and case law. In addition, we argue that the hardcoding of such rules undermines the *moral legitimacy* of systems that implement legal code in this way. The moral legitimacy would be negatively impaired because a system is not flexible nor malleable for a user or to outside circumstances. We will elaborate on this discussion further below (see Section 4 "Softcoding as a Path for More Responsiveness, Flexibility, and Transparency").

In contrast, a "softcoded" solution links executable code with regulation that is expressed - readable for humans as well as machines - in openly accessible documents. This has implications on several levels: Regarding the *architectural design of a software system* (or a cyber-physical system), it means that an explicit effort must be taken to decouple such parameters from the compiled, executable, program. Instead, the system would be designed so that the parameters are loaded, at run time, from a remote source (e.g., a publicly available knowledge base or database), where that remote source needs to be semantically aligned with the system (e.g., through a shared ontology, corresponding to component (1) above). Such a system would then be configured to adapt it to different execution contexts (e.g., different jurisdictions) by swapping this remote source while keeping the same executable code. Finally, *during operation*, the system would retrieve the decision parameters from the configured remote source and thereby adapt its execution (corresponding to components (3) and (4) above) given its context (corresponding to component (2) above). The timeliness and frequency of these retrieval operations here depend on the context and the concrete decisions that the system needs to take - in some situations, it might be sufficient to update the parameters only upon specific trigger events (e.g., a location change) while in other circumstances, regular updates might be required.

2.3 Machine-understandable Data Protection Law

To enable systems that adapt to regulation as outlined above, we first require a way to express law so that it can be interpreted by machines, corresponding to component (1) above; these machine-interpretable documents then form the basis of run-time- adaptations (components (3) and (4)) based on context data (component (2)). For several decades, researchers across the domains of computer science, information systems, and law have been working on representing legal circumstances and documents in a way that would make them automatically interpretable by machines in this way. Setting the stage for such automatic interpretations of legal documents are legal support software that cover simple extensions to text processing systems, collaboration tools for contract drafting (e.g., Beagle),[37] contract highlighting/visualization (e.g., LegalSifter)[38] and term extraction (e.g., LegalRobot)[39]. In addition to these tools, the domain of legal document analytics comprises algorithms that can be run across documents from several data sets and dictionaries and support automatic text analysis and legal text mining.[40] The *ontological* modeling of legal terms and their relationships adds the potential of better structuring and indexing information from legal documents to prepare it for more efficient searching and even for automated reasoning, in addition to providing a foundation to better understand legal terms in their context and for semantic integration[41], e.g., to contrast across (legal) domains or jurisdictions, harmonize documents, and as a bridge between technical and legal perspectives.[42] In this field, lightweight ontologies and taxonomies are used for *describing* concepts and domains while domains can also be *axiomatized* through heavyweight ontologies. This axiomatization creates a foundation for automatic problem-solving, such as fully automatic compliance checking,[43] and such automatic checks have been proposed in the context of complying with specific norms of the GDPR.[44]

To enable automatic compliance checks with the GDPR, systems require access to high-level descriptions of data processing actions (e.g., *storing* or *deletion* of data) and to machine-understandable formalizations of the relevant parts of the underlying legal basis (e.g., GDPR).[45] In addition, the software that performs the processing needs to be (automatically or manually) annotated to allow its interpretation in the context of these formalizations and thereby permit the fusing of legal and program code. A current overview of the state of the art in the domain is given by Rodrigues and his colleagues[46]; in addition, researchers have analyzed the GDPR using formal concept analysis to recover concepts, attributes, and implications with the same level of formality and rigor with which the regulation was created with the goal of supporting more GDPR-consistent systems and service design.[47] While a full axiomatization of legal documents such as the GDPR is currently out of reach,[48] it is, based on such manual analysis, possible to encode *aspects of regulations* that should

37 https://www.capterra.com/p/142807/Beagle/ (accessed 28 October 2020).

38 https://www.legalsifter.com/ (accessed 28 October 2020).

39 Sudhir Agarwal, Kevin Xu and John Moghtader, 'Toward Machine-Understandable Contracts' in *A14J – Artificial Intelligence for Justice* (Workshop at the 22nd European Conference on Artificial Intelligence, The Hague, The Netherlands, August 2016) 5.

40 Charalabidis and others propose a range of applications of such legal text mining including parallel search across legal frameworks that are formulated in different languages, automatic assessment of the degree of transposition of national and international laws (e.g., regarding the relationship of EU Directives and national legislation), comparative analyses of connected laws, timeline analysis including the interrelation of laws and news articles, and text- or even geographically-based visualization. Cf. Yannis Charalabidis and others, 'Use Case Scenarios on Legal Text Mining', in Ben Dhaou Soumaya, Carter Lemuria and Mark A Gregory (eds), *ICEGOV2019: Proceedings of the 12th International Conference on Theory and Practice of Electronic Governance* (Melbourne VIC Australia April 2019, Association for Computing Machinery, 2019) 364.

41 Núria Casellas, *Legal Ontology Engineering* (Springer Netherlands, Dordrecht 2011) 50.

42 Cleyton M d O Rodrigues and others, 'Legal ontologies over time: A systematic mapping study' (2019) 130 *Expert Systems with Applications* 12, 12 et seqq.

43 Rodrigues and others (n 42), 12 et seqq.

44 Piero A Bonatti and others, 'Machine Understandable Policies and GDPR Compliance Checking' (2020) https://arxiv.org/pdf/2001.08930.pdf (accessed 28 October 2020) 1 et seqq.

45 Bonatti and others (n 44), 1 et seqq.

46 Rodrigues and others (n 42), 12 et seqq.

47 Damian A Tamburri, 'Design principles for the General Data Protection Regulation (GDPR): A formal concept analysis and its evaluation' (2020) 91 *Information Systems* 101469.

48 Bonatti and others (n 44), 1 et seqq.

hold unambiguously and without reference to their interpretation contexts. From a technical perspective, such systems thus softcode legal contexts that are described in a transparent way and within openly accessible legal ontologies; and we can even conceive of systems that allow users to modify which of a range of legal (and possibly even personalized) ontologies to use at run time.[49]

Within systems that encode aspects of regulation in this way, one approach towards enabling the automatic processing of contracts, policies, and law is explicit rule-based modeling. These rules are then applied to generate exact legal consequences such as obligations and prohibitions when a specific process execution is identified as part of a monitored workflow.[50] Workflow systems are thereby enabled to initiate actions only after consulting a database with regulatory clauses in order to determine active obligations; the machine-readable representations of clauses and rules however currently need to be created manually. Approaches from the Semantic Web domain, in particular ontologies and vocabularies that are defined using languages from the families of the Resource Description Framework (RDF) and the Web Ontology Language (OWL) can be used for their expressivity and to increase the interoperability of such solutions, while the limits of these standards in the context of conceptualizing the legal domain remain little explored.[51] Legal reasoning is, in principle, defeasible,[52] and it is therefore not possible to decide all juridical nuances using classical logic while formalizing the domain using a monotonic logic only is labor-intensive and might not be understandable by domain experts.[53] Manual encoding of documents by applying non-classical logics may also not scale to a full legal corpus.[54] Moreover, legal rules may conflict with each other, which is resolved through meta-rules that define priority relationships and require defeasible logics.[55] While thus both rule languages (such as LegalRuleML) and languages that correspond to description logics (such as OWL2) have been used as policy languages,[56] policy-reasoning tasks are decidable only in the latter while compliance-checking is undecidable in rule languages, or at least intractable in the absence of recursion.[57]

Researchers have thus been working on the creation of ontologies for the legal domain for several decades with the goals of establishing common and unambiguous terminology and of making the domain accessible to automated processing.[58] Description models of a wide variety of types and on many different abstraction levels have been created. Generally, the manual development of ontologies by knowledge engineers and with the support of domain experts starting from concepts of the target domain is referred to as top-down ontology development and is distinguished from bottom-up approaches where ontologies are extracted by mapping from underlying data sources (e.g., legal documents).[59] In the legal domain, top-down approaches include the *Legal Knowledge Interchange Format* (LKIF) and its *core ontology of basic legal concepts*[60] that is arranged in three clusters: *legal-action*, *legal-role*, and *norm*. To give a concrete example, the norm cluster defines concepts such as *Contract*, *Decree*, and *Treaty*; it then expresses that documents of type *Contract* bear at least one entity of type *Norm* that are held by agents of type *Natural_Person* or *Legal_Person* towards some *Thing* (e.g., an action) that is normatively qualified (i.e., allowed or disallowed).[61]

For applying such an ontology in a practical application, it needs to be complemented with a more specific legal domain ontology and with a formalization and vocabulary of the underlying argumentation and reasoning which represents the structure and dynamics of argumentation that shall be applied.[62] In other words, these models are typically only loosely coupled with the actual legislation text which makes it difficult to verify whether they are effective[63] and accurate with respect to their representation of law. Consequently, there is a lack of practical adoption and the body of academic work is criticized, for instance regarding specific omissions that constrain practical usage.[64] Together with the challenges around the rule-based modeling of the legal domain discussed above, there has thus also not been an instantiation of LKIF and LegalRuleML at scale or used for formalizing or annotating the content of a legal corpora either automatically or manually.[65] To overcome this gap between research and practice, recent work targets the design of semantic systems that can be used to express legal circumstances in *specific domains* (e.g., to express legislative obligations[66]) and often coupled to *specific use cases*. Only then are these connected to more abstract knowledge models—in the case of [67] as an extension profile that can be used to model obligations with the *Open Digital Rights Language* (ODRL).[68] While the design of such extensions is thus from the beginning informed from approaches such as ODRL and LKIF, the implementation is done in a bottom-up way, and the combined system is in addition instantiated in the form of a usable tool.[69]

3. Encoding Data Protection Law: An Imperfect Remedy

Unsurprisingly, the increased deployment of smart devices like social robots has led to an increased interest among academics in

49 Kimberly Garcia and others (n 10).
50 Alan Abrahams, David Eyers and Jean Bacon, 'An asynchronous rule-based approach for business process automation using obligations' in Bernd Fischer (ed.), *Proceedings of the 2002 ACM SIGPLAN workshop on Rule-based programming* (ACM, New York, NY 2002).
51 Rodrigues and others (n 42), 12 et seqq.
52 Juan B Carlos, 'Why is Legal Reasoning Defeasible?' in Arend Soeteman (ed.), *Pluralism and Law* (Springer, Dordrecht 2001).
53 Rodrigues and others (n 42), 12 et seqq.
54 Guido Governatori and others, 'Norm Modifications in Defeasible Logic' in Marie-Francine Moens and Peter Spyns (eds), *Legal Knowledge and Information Systems, JURIX 2005: Eighteenth Annual Conference* (IOS Press 2005) 13 et seqq.
55 Marcello Ceci, 'Combining Ontologies and Rules to Model Judicial Interpretation' in *Proceedings of the RuleML@ECAI 6th international doctoral consortium* (Montpellier, France, August 2012) 2.
56 Bonatti and others (n 44), 1 et seqq.
57 Bonatti and others (n 44), 1 et seqq.; Piero A Bonatti, 'Datalog for Security, Privacy and Trust' in Oege de Moor and others (eds), *Datalog reloaded: First international workshop, Datalog 2010, Oxford, UK, March 16 - 19, 2010, revised selected papers* (Lecture Notes in Computer Science vol 6702. Springer 2011).
58 Rodrigues and others (n 42), 12 et seqq.
59 Biralatei Fawei and others, 'A Semi-automated Ontology Construction for Legal Question Answering' (2019) 37 *New Gener. Comput.* 453.
60 Rinke Hoekstra and others, 'The LKIF Core Ontology of Basic Legal Concepts' in Pompeu Casanovas and others (eds), *Proceedings of the 2nd Workshop on Legal Ontologies and Artificial Intelligence Techniques* (Stanford, CA, USA 2007) 43 et seqq.
61 The LKIF core ontology is available at https://github.com/RinkeHoekstra/lkif-core (accessed 28 October 2020).
62 Ceci (n 55), 2.
63 Sushant Agarwal and others, 'Legislative Compliance Assessment: Framework, Model and GDPR Instantiation' in Manel Medina and others (eds), *Privacy Technologies and Policy: 6th Annual Privacy Forum, APF 2018, Barcelona, Spain, June 13-14, 2018, Revised Selected Papers* (Security and Cryptology vol 11079, Springer International Publishing 2018) 131 et seqq.
64 Agarwal and others (n 63), 131 et seqq.
65 Fawei and others (n 59), 453 et seqq.
66 Agarwal and others (n 63), 131 et seqq.
67 Agarwal and others (n 63), 131 et seqq.
68 https://www.w3.org/TR/odrl/ (accessed 28 October 2020).
69 Cf. Agarwal and others (n 63), 131 et seqq. for GDPR compliance assessment.

the interaction between regulation and social robots.[70] Leenes and Lucivero differentiate between four scenarios: First, the ability of *law to regulate the design of a robot*, second, the *ability of a robot to regulate user behavior* through its design, third, the *ability of law to regulate the effects of a robot's behavior*, and fourth, the *ability of code to regulate a robot's behavior*.[71] Encoding data protection as enshrined in the GDPR focuses in particular on the first and last category mentioned by Leenes and Lucivero: Ensuring that the external and internal design of a robot complies automatically with the fundamental principles of data protection law (e.g., transparency about the data gathering, limitation of data processing practices, deactivation of functionality upon lacking user consent). Thereby, encoding data protection regulates the potential privacy implications and effects of a social robot and thus the impact this robot has on user behavior (e.g., a privacy-friendly robot might increase user comfort, while a privacy-invasive one may lead to chilling behaviors). As mentioned above (see Section 2.2 "Hard- or Softcoding Law") the design process ideally will not only lead to configuring robots with the data protection principles in mind but also constructing devices that at run time can adapt to contextual changes.

As described in Section 2 "From an Ideal to Implementations", while remedies to encode data protection have been proposed, they have encountered various obstacles. In the following, we map the issues that arose in the implementation of the data protection principles in a social robot[72] and refer to other research projects and literature highlighting similar difficulties.[73] While our findings stem from an investigation on the implementation of data protection by design and thus focus on data protection law, they apply to legal code across legal domains. In fact, different examples[74] of encoding of law can be found which show that, depending on the characteristics of the legislation at hand (e.g., ones involving calculations, relying on machine-readable factual information, involving compliance with processes),[75] the difficulties arising in implementing the law into the design vary (see Section 4 "Softcoding as a Path for More Responsiveness, Flexibility, and Transparency"). The difficulties arise in particular when dealing with balancing norms rather than procedural ones (or muddy norms instead of crystal norms[76]). Former norms are more vague and open to interpretation. Here we see difficulties that arise from the need to come up with assumptions (e.g., de facto hierarchies), 'solve' conflicts within the law, determining how to deal with balancing tests and legitimacy criteria, generalize legal terms to encode them, and disentangle connected norms. Moreover, the lack of automatic access to machine-readable documentation and the difficulties of assessing risk complicate the implementation of law into code. Lastly, we discuss the business implications and potential constraints to encoding data protection principles.

3.1 Encoding Assumptions

Encoding data protection implies coming up with solutions when the law is silent, vague, and ambiguous.[77] Doing so requires relying on assumptions, even when those may be well founded and documented in the literature. In that sense, law indulges in the luxury (and, sometimes, necessity) of staying vague, but code cannot.[78] Nonetheless, if no clear case law in favor of one or the other interpretation exists in a general manner, even the most well-argued assumption remains debatable and defeasible. One example that illustrates this difficulty arises when encoding the principle of lawfulness: The purpose of the processing determines the legal ground, which in turn must be established before the processing occurs. Thus already the choice of the legal ground becomes dependent on other characteristics of the processing that are determined at the design stage. In addition, as will be explained below, since no hierarchy of legal grounds can be found within the law or case law, developers will be motivated to create a de-facto normative hierarchy, which ultimately is subjective and imposed by system designers and engineers.

According to the Article 29 Working Party (WP29), the data controller must *determine which lawfulness ground is the most appropriate in a given scenario*. Not all the processing can thus be justified by consent but only instances in which consent is the appropriate lawfulness ground. This provision by the WP29 has been criticized.[79] But case law has made clear that the choice of the appropriate legal basis is key and an *inappropriate ground for processing leads to fines and inability to claim other legal grounds at a later point of time*.[80] One could interpret the WP29 opinion and the cited case law as such that if other lawfulness grounds than consent are applicable, those need to be given priority in the design and implementation process. In other words, a data controller needs to first check whether data can be processed on other legal grounds than consent given its current context, and if that is not the case require consent of the data subject. But of course, such an interpretation is highly controversial,[81] and depending

70 Leenes and Lucivero (n 11), 198; Bibi van den Berg. 'Robots as Tools for Techno-Regulation' (2011) 3 *Law, Innovation and Technology* 319; Christoph Lutz and Aurelia Tamò, 'RoboCode-Ethicists' in *Proceedings of the 2015 ACM Web Science Conference* (Oxford, United Kingdom, June – July 2015).

71 Leenes and Lucivero (n 11), 198.

72 Kimberly Garcia and others (n 10).

73 Leenes and others (n 5); Koops and Leenes (n 2), 159; Leenes and Lucivero (n 11), 193.

74 Cf. for examples e.g., the OECD Working Papers on Public Governance, 'Cracking the code: Rulemaking for humans and machines' (2020) available at https://www.oecd-ilibrary.org/governance/cracking-the-code_3afe6ba5-en (last accessed 20 December 2020).

75 Cf. findings of New Zeland LabPlus in 2018 https://www.digital.govt.nz/dmsdocument/95-better-rules-for-government-discovery-report/html (accessed 8 November 2020).

76 A term coined by Carol M Rose, 'Crystals and Mud in Property Law' (1988) 40 *Stanford Law Review* 577.

77 Cf. Leenes and others (n 5), 28 elaborating on the vagueness, open texture, and ambiguity of law.; cf. also on delineating the scope of data requirements Koops and Leenes (n 2), 163.

78 We note that the law often remains vague for good reasons; we do not mean to disesteem these reasons, but note that the vagueness creates an obstacle to the encoding of law.

79 Winfried Veil, 'Einwilligung oder berechtigtes Interesse? – Datenverarbeitung zwischen Skylla und Charybdis' (2018) 71(46) Neue Juristische Wochenschrift 3337, 3338.

80 EDPB, 'Company fined 150,000 euros for infringements of the GDPR' (31 July 2019) https://edpb.europa.eu/news/national-news/2019/company-fined-150000-euros-infringements-gdpr_en (accessed 28 October 2020) Hellenic DPA fines PWC reason is that the company asked for consent for the processing of data, yet this was seen as an inappropriate legal ground as the processing was covered by another legal ground that was not mentioned to the employees. This decision shows that reversing the legal ground is not readily possible, as the infringement has consequences with respect to the data that has been processed without appropriate legal ground.

81 Even the WP29 seems to have contradicting options on said matter. Cf. Article 29 Working Party, 'Guidelines on consent under Regulation 2016/679' (WP 259 rev.01, 10 April 2018), at 3 in conjunction with 23 and Article 29 Working Party, 'Opinion 8/2014 on the on Recent Developments on the Internet of Things' (WP 223, 16 September 2014), at 15 where the WP29 states that "Consent (Article 7(a)) is the first legal basis that should be principally relied on in the context of the IoT, whether by device manufacturers, social or data platforms, devices lenders or third party developers". The contradiction between the two opinions has not been addressed in the literature. The recent case of the Hellenic DPA (see above footnote 80) shows however clearly that appropriate lawfulness grounds are necessary.

on the interpretation (and national understanding of data protection law as a whole), different approaches could be proposed.[82] One argument to give priority to other legal grounds prior to resulting to consent is the following: Both consent and data processing necessary for the performance of a contract are based on the idea that a user/data subject gives consent to a specific action or manifests an intent to enter into a (contractual) relationship with the data controller. Yet, in particular consent is inherently linked with problems with respect to its efficacy to provide control over data processing.[83] Thus, legal grounds that are not affected (as much) by cognitive biases discussed in the literature shall be given priority. These grounds are based on a legislative process or have been established by case law. In any case these grounds are tied to a democratically established process, which arguably should give them more weight. That being said, the resulting engineering implications are to *determine a hierarchy for testing legal grounds* (e.g., (1) processing based on a legal obligation; (2) processing based on legitimate interests (3) processing necessary for the performance of a contract, and (4) processing based on consent). While such an interpretation enables taking the purpose into account (e.g., in case of processing of data in an employment situation to pay benefits to employee, the first legal ground in the hierarchy could be fulfilled; e.g., in the case of processing for marketing purposes, the fourth legal ground would be fulfilled) to automatically test for a legal ground, the result in practice might be that the de-facto hierarchy set by developers will lead to relying on an inadequate legal ground, as a decision must be taken in order to proceed.

This obstacle sheds light on a difficulty that often arises when trying to embed data protection into the design: The vagueness of the law and potential syntactic ambiguity complicates and potentially impedes such endeavours. In our opinion, while vagueness is acceptable when dealing with balancing tests and legitimacy criteria within the principles, determining the adequate legal ground has a procedural element to it which projects some sort of hierarchy and thus requires more specific guidance – not only for engineers but also for lawyers. In the end, we see here how European data protection law, which in itself is a compromise between different approaches in the member states, fails to reconcile these different approaches to its fullest, which become apparent when trying to embed data protection into the design.

3.2 'Solving' Conflicts in the Law

In addition, *tensions between different principles* need to be addressed and practically feasible processes of how to solve those tensions need to be devised when designing for compliance with data protection law.[84] Clear mechanisms on how to resolve such have not been widely discussed in the literature yet are necessary in order to determine the legal code implications thereof.

In particular, a conflict between the principle of accuracy and data minimization has been raised in the literature.[85] The principle of accuracy aligns with the interests of the data controller, who has an interest in having accurate and up-to-date data.[86] The principle of accuracy is also linked to data security by means of requiring the integrity of the data (i.e., that the data is maintained as it was originally collected) as well as its trueness and veracity.[87] However, even originally correct data that has not been changed can become inaccurate after a certain time has elapsed, as the principle of accuracy is context-dependent.[88] In fact, the principle of accuracy exists not as a stand-alone principle, but as a connecting principle. The 'connecting' aspect of accuracy can for instance be seen in the ECJ's Google vs. Spain decision that ultimately links accuracy of data to the fairness principle, by stating that out-of-context information can lead to unfair decisions or judgements.[89] By that token though, the principle of accuracy does not seem to be much in conflict with the principle of data minimization (which in turn is interlinked with the principle of purpose and storage limitation).[90] A core design feature under the GDPR is to process only the (minimum) data necessary to achieve a specified purpose. This implies also to limit the storage of the data to only that data that is necessary to achieve said purposes. These principles set the data controller under pressure to be able to justify why certain data is being collected, processed, and kept, and thereby strongly decreases the data controller's incentives to keep unnecessary data. In fact, from a technical perspective the principle of accuracy and data minimization can be encoded, for instance by implementing expiration dates on data processing operations.[91]

Another aspect that conflicts with the data minimization principles is the fact that the controller bears the burden of proof that valid consent was obtained when relying upon that legal ground.[92] This

82 The German literature seems to typically praise consent as the ultimate means to establish informational self-determination. Cf. e.g., Marie-Theres Tinnefeld and Isabell Conrad, 'Die selbstbestimmte Einwilligung im europäischen Recht' (2018) 9 *Zeitschrift für Datenschutz* 391, 392; Dirk Heckmann and Anne Paschke, 'Art. 7 Bedingungen für die Einwilligung' in Eugen Ehmann and Martin Selmayr (eds), *DS-GVO: Datenschutz-Grundverordnung: Kommentar* (2nd edn, Beck'sche Kurz-Kommentare, C.H. Beck, LexisNexis 2018) 9. However, other scholars from Germany seem to have a more critical stance, cf. Stefan Ernst, 'Die Einwilligung nach der Datenschutzgrundverordnung' (2017) 3 *Zeitschrift für Datenschutz* 110, 110.

83 Chris J Hoofnagle, Bart van der Sloot and Frederik Z Borgesius, 'The European Union general data protection regulation: what it is and what it means' (2019) 28(1) *Information & Communications Technology Law* 65, 80; Daniel J Solove, 'Privacy Self-Management and the Consent Dilemma' (2013) 126 *Harv. L. Rev.* 1880, 1883 et seqq.; cf. Elettra Bietti, 'Consent as a Free Pass: Platform Power and the Limits of the Informational Turn' (2020) 40(1) *Pace Law Review* 310.

84 Koops and Leenes (n 2), 166 et seq.; Leenes and Lucivero (n 11), 211 et seqq.

85 Cf. Erik Zouave and Jessica Schroers, 'You've been Measured, You've been Weighed & You've been Found Suspicious - Biometrics & Data Protection in Criminal Justice Processing' in Ronald Leenes, Rosamunde van Brakel and Serge Gutwirth (eds), *Data protection and privacy: The Internet of Bodies* (Computers, privacy and data protection 2018) 9; cf. Pagallo (n 2), 343; cf. Michael Veale, Reuben Binns and Jef Ausloos, 'When data protection by design and data subject rights clash' (2018) 8(2) *International Data Privacy Law* 105.

86 Thomas Hoeren, 'Big Data and Datenqualität – ein Blick auf die DSGVO' (2016) 10 *Zeitschrift für Datenschutz* 459; Gloria Gonzáles Fuster, 'Inaccuracy as a privacy-enhancing tool' (2010) 12(1) *Ethics of Information Technologies* 87; the principle of accuracy is also a guiding principle in the OECD 1980 and now 2013 Guidelines.

87 Hoeren, (n 86), 459 with reference to the ISO Standard 5725-1: 1994.

88 Cf. Google Spain SL and Google Inc. v Agencia Española de Protección de Datos (AEPD) and Mario Costeja González, Case C-131/12, [2014] (ECLI:EU:C:2014:317), at para. 93.

89 Cf. Google Spain SL and Google Inc. v Agencia Española de Protección de Datos (AEPD) and Mario Costeja González, Case C-131/12, [2014] (ECLI:EU:C:2014:317); Rolf H. Weber and Simon Henseler, 'Regulierung von Algorithmen in der EU und in der Schweiz: Überlegungen zu ausgewählten Regulierungsthemen' (2020) 28 *Zeitschrift für Europarecht* 31.

90 Data minimization relies on the purpose for which the data is being processed as it requires that only data that is absolutely necessary to achieve said purpose is being processed; storage limitation can be seen as a form of data minimization as it requires erasing data that is no longer necessary for achieving a stated purpose.

91 Bart Custers, 'Click here to consent forever: Expiry dates for informed consent' (2016) 3(1) *Big Data & Society* 1.

92 Cf. Art. 7(1) GDPR. EDPB, 'Guidelines 05/2020 on consent under Regulation 2016/679 Version 1.1' (4 May 2020), at 22; Article 29 Working Party,

requires data controllers to 'store the declaration of consent together with the name of the data subject or another reliable identifier (email address, etc.) and the time of the consent ('timestamp')' as long as the processing activity persists.[93] Moreover, whenever data controllers target children (e.g., connected toys), the controller must ensure that parental consent is obtained when the data subject is below a certain threshold.[94] Although the GDPR does not demand the controller to verify the age of the child, it might be inevitable in practice, given that mechanisms for age confirmation can easily be circumvented.[95] Different age verification mechanisms exist, yet it is likely that all of them put at risk the privacy of children by requiring the collection of additional personal data.[96] Moreover, where a device processes data continuously or periodically, it is possible that during this time the child may exceed the age threshold and parental consent is no longer required, but the consent of the child himself. Since relying on parental consent after the child has reached the respective age makes the processing unlawful, the data controller is likely to record not only the declaration of consent together with the name of the data subject or another reliable identifier (as seen above), but also the child's date of birth, in order to ensure that the system can obtain the child's own consent once the child reaches the respective age threshold.[97]

Another conflict in the law can be found in the prospective element of transparency. The wording of Article 22 of the GDPR stipulates a right of the data subject to object to specific forms of automated decision-making practices, yet the article prohibits such practices unless explicit consent is provided. Unsurprisingly, this has triggered a debate on whether it qualifies as a right or as a prohibition all together. This conflict remains unresolved in the literature, as arguments in favor of a right[98] and in favor or a prohibition[99] can be found. Also a historical analysis cannot resolve this conflict, as some member states had interpreted the former Article 15(1) DPD as a prohibition (while others had not).[100] While the WP29 - and the EDPD - however seem to agree that despite the wording as a right Art. 22(1) of the GDPR and its position within Chapter III of the GDPR should qualify as a prohibition,[101] also arguments in favor of an individual right are abundant. Especially, the fact that the information duties of data controllers listed in Articles 12 to 14 of the GDPR include a requirement to mention if automated decision-making occurs (which would not be needed if no such decision-making would be allowed) point towards an individual right.[102]

3.3 Dealing with Legitimacy

Another difficulty arises whenever the law refers to legitimacy criteria and balancing of competing interests. An example thereof is determining when the legal ground of *legitimate interests*, which is only applicable in the context of businesses and users,[103] can be applied. Data controllers may argue – in line with the WP29 statement – that sometimes they 'temporarily need to perform some facial recognition processing steps precisely for the purpose of assessing whether a user has provided consent or not as a legal basis for the processing. This initial processing (i.e., image acquisition, face detection, comparison, etc.) may in that case have a separate legal basis, notably the legitimate interest of the data controller to comply with data protection rules. Data processed during these stages must only be used for the strictly limited purpose to verify the user's consent and should therefore be deleted immediately after.'[104] But this statement does not exempt from an assessment of the reasonable expectations of a data subject at the time of the collection which is based on the relationship with the controller.[105] The reasonable expectation relates to the 'foreseeability and acceptance from the side of the data subject of the processing operation. While the foreseeability needs to be articulated objectively (clear, timely, and transparent information notice, justified for the purposes it serves, etc.) by the data controller, the acceptance of the data subject can also be implied (otherwise, we would refer to 'consent').'[106]

'Guidelines on consent under Regulation 2016/679' (n 81), at 20.

93 Sebastian Dienst, 'Lawful processing of personal data in companies under the General Data Protection Regulation' in Tobias Kugler and Daniel Rücker (eds), *New European general data protection regulation: A practitioner's guide: ensuring compliant corporate practice* (C.H. Beck; Nomos; Hart 2018) 99; cf. EDPB, 'Guidelines 05/2020 on consent under Regulation 2016/679 Version 1.1' (n 92), at 23.

94 Cf. Art. 8(1) GDPR.

95 Article 29 Working Party, 'Guidelines on consent under Regulation 2016/679' (n 81), at 25 et seq.; Eleni Kosta, 'Article 8. Conditions applicable to child's consent in relation to information society services', *The EU general data protection regulation (GDPR): A commentary* (OUP 2020) 360 et seqq.

96 Unicef, 'Children's Online Privacy and Freedom of Expression' (May 2018), https://issuu.com/unicefusa/docs/unicef_toolkit_privacy_expression?e=29613278/60947364 (accessed 29 October 2020) 15; cf. Article 29 Working Party, 'Guidelines on consent under Regulation 2016/679' (n 81), at 27.

97 Cf. Koops and Leenes (n 2), 165 with respect to the Dutch implementation of the DPD; cf. also Kosta, 'Article 8. Conditions applicable to child's consent in relation to information society services' (n 95), 361 et seq.

98 Wulf Kamlah, 'Art. 22 DSGVO' in Kai-Uwe Plath (ed.), *DSGVO/BDSG Kommentar* (3rd edn, Dr. Otto Schmidt 2018) 4; Anton Vedder and Laurens Naudts, 'Accountability for the use of algorithms in a big data environment' (2017) 31(2) *International Review of Law, Computers & Technology* 206, 213 et seq.

99 Cf. e.g., Isak Mendoza and Lee Bygrave, 'The Right Not to Be Subject to Automated Decisions Based on Profiling' in Tatiani Synodinou and others (eds), *EU Internet Law: Regulation and Enforcement* (Springer 2017), 86 et seq.; Lee Bygrave, 'Minding the Machine v.2.0: The EU General Data Protection Regulation and Automated Decision Making' in Karen Yeung and Martin Lodge (eds), *Algorithmic Regulation* (Oxford University Press 2019) 246; Frederike Kaltheuner and Elettra Bietti, 'Data is power: Towards additional guidance on profiling and automated decision-making in the GDPR' (2018) 2(2) *Journal of Information Rights, Policy and Practice* 1, 10 et seq.; Eike Mario Martini, 'Art. 22. Automatisierte Entscheidungen im Einzelfall einschließlich Profiling' in Boris P Paal and Daniel A Pauly (eds), *Datenschutz-Grundverordnung, Bundesdatenschutzgesetz* (Beck'sche Kompakt-Kommentare, 2nd ed. C.H.Beck 2018) 29; Guido Noto la Diega, 'Against the Dehumanisation of Decision-Making – Algorithmic Decisions at the Crossroads of Intellectual Property, Data Protection, and Freedom of Information' (2018) 9(1) *JIPITEC* 3, 17.

100 Bygrave, 'Minding the Machine v.2.0: The EU General Data Protection Regulation and Automated Decision Making' (n 99), 6.

101 Cf. Article 29 Working Party, 'Guidelines on Automated individual decision-making and Profiling for the purposes of Regulation 2016/679' (WP251 rev.01, February 2018), at 19. Potentially, Article 11 of the GDPR, which exempts data controllers from having to comply with individual rights but excludes Article 22 from this exemption indicates thereby that a difference between individual rights (Art. 15-20 GDPR) and Art. 22 of the GDPR exists. This could be taken to mean that Art. 22 has to be treated differently from individual rights. Cf. also Maja Brkan, 'Do algorithms rule the world? Algorithmic decision-making and data protection in the framework of the GDPR and beyond' (2019) 27 *International Journal of Law and Information Technology* 91, 99.

102 Lee Bygrave, 'Article 22. Automated individual decision-making, including profiling', *The EU general data protection regulation (GDPR): A commentary* (OUP 2020) 531.

103 Rec. 47 excluding the applicability of this legal ground in the case of state and citizens.

104 Article 29 Working Party, 'Opinion 02/2012 on facial recognition in online and mobile services' (WP 192, 22 March 2012), at 5.

105 Cf. Rec. 47; Irene Kamara and Paul de Hert, 'Understanding the Balancing Act Behind the Legitimate Interest of the Controller Ground: A Pragmatic Approach' (Brussels Privacy Hub Working Paper, August 2018) https://brusselsprivacyhub.eu/BPH-Working-Paper-VOL4-N12.pdf (accessed 28 October 2020) 10.

106 Kamara and de Hert (n 105), 17.

Determining when processing can be based on legitimate interests, and taking the reasonable expectations into account, is not a trivial task; and encoding of this process is even more complicated. In fact, to do so requires conducting three tests that are interlinked with one another and lead to an overall balance of interests. To put it differently, the balance of interests test, which is the final step out of three, necessitates two prior steps: a legitimacy of interests test and adequacy test.[107] The *legitimacy test* requires a proof of a legitimate interest by the data controller.[108] According to the WP29, legitimate interests of data controllers must be real and present interests that the data controller has articulated. In other words, future interests, i.e., ones that depend on the fulfilment of a future condition or expectation, are not sufficient. The WP29 also notes that the 'concept of 'interest' is closely related to, but distinct from, the concept of 'purpose'.[109] While a purpose relates to any aim of the data processing, the interests relate to the broader stake the controller has in the processing and the benefit the controller derives from that processing.[110] An interest is not considered to be legitimate 'where the processing is not genuinely necessary for the performance of a contract but rather relates to the ancillary use of data and is achieved through terms unilaterally imposed on the data subject.'[111] The GDPR mentions examples of legitimate interests such as preventing fraud and direct marketing[112] and ensuring network and information security.[113] Those interests are likewise mentioned by the WP29.[114] In case law, different legitimate interests have emerged: In Case C-708/18[115] in which the court had to determine the legitimacy of installed video surveillance in the common parts of a building, the court weighed the interests in the protection of the property and the health and life of co-workers against the right to privacy. The court saw the data processing as legitimate as it argued that the data controller had no other means available that were less invasive to ensure the mentioned interests. In a similar case[116] the court followed the same argument. In another decision,[117] the court acknowledged that the interests of 'a third party in obtaining the personal information of a person who damaged their property in order to sue that person for damages can be qualified as a legitimate interest.'[118] In a recent case, a Dutch court overturned the Dutch DPA's restrictive interpretation of Article 6(1)(f), according to which commercial interests cannot be legitimate interests. Following the European Data Protection Board's guidelines, the court instead found that purely commercial interests are legitimate interests, provided they are real and not speculative.[119] In contrast to those cases acknowledging a legitimate interest as a legal ground, in its famous *right to be forgotten* decision, the ECJ argued that purely economic interests of the search engine provider in not de-indexing certain information are not legitimate interests.[120] The *adequacy test* looks at whether the processing is indeed necessary to achieve the interests or if less intrusive means would be available.[121] The case law above also illustrates how adequacy/necessity are context- or case-dependent. Lastly, the *balancing test* takes into account the impact of the data processing on the data subject.[122] This requires an assessment that takes the positive and negative (potential) consequences into account.[123] While positive consequences can include the interests of the data controller, those interests can overlap with those of the broader community (e.g., freedom to conduct business, of information, science). Negative consequences include potential adverse effects such as emotional impacts and chilling effects.[124] As such emotional and behavioral impacts are difficult to predict, caution should be employed when arguing such consequences let alone codifying them. Nonetheless, based upon the WP29 Opinion on legitimate interests,[125] some criteria are mentioned that more likely tip the balance in one direction or the other: For instance, if sensitive data such as biometric data is being processed, more severe negative consequences are assumed,[126] likewise in case of data being 'publicly disclosed or otherwise made accessible to a large number of persons, or whether large amounts of personal data are processed or combined with other data.'[127]

The question of legitimacy does not only arise with respect to finding

107 Cf. also Autorité de protection des données, 'Recommandation n°01/2020 du 17 janvier 2020 relative aux traitements de données à caractère personnel à des fins de marketing direct' (17 January 2020) https://www.autoriteprotectiondonnees.be/publications/recommandation-n-01-2020.pdf (accessed 28 October 2020) on these three tests.
108 Kamara and de Hert (n 105), 12.
109 Article 29 Working Party, 'Opinion 06/2014 on the notion of legitimate interests of the data controller under Article 7 of Directive 95/46/EC' (WP 217, 9 April 2014), at 24.
110 Article 29 Working Party, 'Opinion 06/2014 on the notion of legitimate interests of the data controller under Article 7 of Directive 95/46/EC' (n 109), at 24.
111 Róisín Á Costello, 'The Impacts of AdTech on Privacy Rights and the Rule of Law' (2020) *Technology and Regulation* 11, 17 with reference to Article 29 Working Party, 'Opinion 06/2014 on the notion of legitimate interests of the data controller under Article 7 of Directive 95/46/EC' (n 109), 16.
112 As mentioned in Rec. 47 GDPR.
113 Rec. 49 GDPR.
114 Article 29 Working Party, 'Opinion 06/2014 on the notion of legitimate interests of the data controller under Article 7 of Directive 95/46/EC' (n 109), at 25 stating: "conventional direct marketing and other forms of marketing or advertisement", "unsolicited non-commercial messages, including for political campaigns or charitable fundraising", "prevention of fraud, misuse of services, or money laundering", "physical security, IT and network security", or "processing for research purposes (including marketing research)".
115 *TK v Asocia ia de Proprietari bloc M5A-ScaraA*, Case C-708/18 [2019] (ECLI:EU:C:2019:1064).
116 *František Ryneš v Ú ad pro ochranu osobních údaj*, Case C-212/13 [2014] (ECLI:EU:C:2014:2428), at para. 34.
117 *Valsts policijas R gas re iona p rvaldes K rt bas policijas p rvalde v R gas pašvald bas SIA 'R gas satiksme'*, Case C-13/16 [2017] (ECLI:EU:C:2017:336), at para. 30 and the case-law cited.
118 *Productores de Música de España (Promusicae) v Telefónica de España SAU, Promusicae*, Case C-275/06 [2008] (ECLI:EU:C:2008:54), at para. 53.
119 Hunton Andrews Kurth LLP's Privacy and Cybersecurity practice, 'Dutch Court Overturns DPA Fine on Legitimate Interest Legal Basis' (1 December 2020) https://www.huntonprivacyblog.com/2020/12/01/dutch-court-overturns-dpa-fine-on-legitimate-interest-legal-basis/ (accessed 21 December 2020); Ady Nieuwenhuizen, 'Judge overturns Dutch DPA GDPR fine' (26 November 2020) https://www.fieldfisher.com/en/insights/judge-overturns-dutch-dpa-gdpr-fine (accessed 21 December 2020).
120 *Google Spain SL, Google Inc. v Agencia Española de Protección de Datos (AEPD), Mario Costeja González*, Case C-131/12 [2014] (ECLI:EU:C:2014:317), at para. 81.
121 Cf. Rec. 39 GDPR.
122 Article 29 Working Party, 'Opinion 06/2014 on the notion of legitimate interests of the data controller under Article 7 of Directive 95/46/EC' (n 109), at 36 et seq.
123 Article 29 Working Party, 'Opinion 06/2014 on the notion of legitimate interests of the data controller under Article 7 of Directive 95/46/EC' (n 109), at 37.
124 Article 29 Working Party, 'Opinion 06/2014 on the notion of legitimate interests of the data controller under Article 7 of Directive 95/46/EC' (n 109), at 37; Moritz Büchi and others, 'The chilling effects of algorithmic profiling: Mapping the issues' (2020) 36 *Computer Law & Security Review* 105367.
125 Article 29 Working Party, 'Opinion 06/2014 on the notion of legitimate interests of the data controller under Article 7 of Directive 95/46/EC' (n 109).
126 Article 29 Working Party, 'Opinion 06/2014 on the notion of legitimate interests of the data controller under Article 7 of Directive 95/46/EC' (n 109), at 38-39.
127 Article 29 Working Party, 'Opinion 06/2014 on the notion of legitimate interests of the data controller under Article 7 of Directive 95/46/EC' (n 109), at 39.

the adequate legal ground but is a question that is at the core of data protection law. In particular, the principle of purpose limitation states that each processing of data must occur for legitimate purposes. It could be argued that the purposes of processing are legitimate if the processing is lawful according to Article 6 of the GDPR.[128] While this seems reasonable for processing that occurs for the purpose of complying with a legal obligation or to protect vital interests, making the legitimacy of a purpose depending on consent seems less reasonable. In particular because of the failings noted in the literature with respect to consent (e.g., failures with respect to accepting terms that are not read, biases of individuals and inability to calculate long-term risks vs. short-term benefits, others). These failures show that the term *legitimate* must likely be understood more broadly, as *in accordance with the law*. According to the WP29 it should include not only primary and secondary legislation but all forms of written law, principles, and case law.[129] In addition, also codes of conduct and ethics and 'the general context and facts of the case' as well as social and technical changes must be taken into account if they affect the legitimacy of a given purpose over time.[130]

3.4 Generalizing Legal Terms

Many aspects encountered within the data protection law cannot today be expressed in a machine-readable way, meaning that depending on the principle at hand case-by-case considerations are key. This is also true for principles for which there is a rich (and evolving) case law and which ultimately require updates as to the factors that courts took into consideration. This results in decisions that are based on the (partially subjective) weighing of different options, and can lead to (un)intended *generalizations* and delineations.[131]

An example thereof is the interpretation of the term 'fairness'. Fairness, transparency, and lawfulness are all closely linked to one another. This link is already apparent in Article 5(1)(a) of the GDPR which ties the concepts together. In other words, formally speaking the concept of fairness can be seen as the middle ground on a spectrum between lawfulness and transparency, providing a link between both concepts. As such a *middleman*, the ideal of fairness is linked to the concept of lawfulness when fairness reflects procedural fairness; and linked to the concept of transparency when fairness reflects 'fairly transparent' processing. Aside from this, fairness in itself must also be understood as 'effect-based' wanting to mitigate imbalances that lead to vulnerability and discrimination.[132]

Understanding fairness as more aligned with lawfulness means to interpret it as *procedural fairness*.[133] This procedural fairness implies a balanced approach with respect to weighing competing interests against each other. What speaks in favor of understanding fairness as procedural fairness is that in some translations of the GDPR in languages of EU member states the term 'fairness' is translated as a term relating closer to lawfulness.[134] On the one hand, fair balancing means taking the context into account in order to prevent unjust 'outcomes' or 'impacts.' On the other hand, procedural fairness requires implementing guiding procedural rules.

The GDPR refers in numerous articles and recitals to 'fair and transparent' processing.[135] This demonstrates the strong link among fairness and transparency and is linked to the information duties as the data subject must have actual knowledge of the main characteristics of the processing of his or her personal data.[136] While the ECJ has interpreted the concept of fairness as a sort of requirement of transparency in the case of the processing of personal data when public authorities transfer data to other authorities,[137] such an interpretation is also possible within the private sector. As Clifford and Ausloos conclude, the court's reasoning in these cases was to provide protection against asymmetric relationships, even in cases where the sharing of data is not malevolent (i.e., instances in which the controller is not trying to deceive a data subject).[138] Interestingly, the ICO and CNIL likewise understand the term 'fairness' as a means to rebalance asymmetric relationships, among others by means of providing more transparency about the underlying data processing.[139]

While aligning the meaning of fairness with lawfulness and transparency would mean with respect to the engineering implications that the provisions of lawfulness and transparency would need to be followed through (with all mentioned caveats), the term fairness

128 Whether or not one agrees with this argument will depend also on whether the term 'lawfulness' is understood broadly or narrowly. Cf. footnote 24 above.
129 Article 29 Working Party, 'Opinion 03/2013 on purpose limitation' (WP 203, 2 April 2013), at 20.
130 Article 29 Working Party, 'Opinion 03/2013 on purpose limitation' (n 129), at 20.
131 Cf. here Koops and Leenes (n 2), 163.
132 Gianclaudio Malgieri, 'The Concept of Fairness in the GDPR: A linguistic and contextual interpretation' in Mireille Hildebrandt and others (eds), FAT* '20: Proceedings of the 2020 Conference on Fairness, Accountability, and Transparency (Barcelona Spain January 2020, Association for Computing Machinery New York, NY, United States). Note that the link between fairness and non-discrimination can already be found within the Resolutions of the Council of Europe on the protection of privacy in electronic data banks from 1973 and 1974, cf. Council of Europe, Committee of Ministers, Resolution 73 (22) on the protection of privacy of individuals vis a vis electronic data banks in the private sector; Council of Europe, Committee of Ministers, Resolution 74 (29) on the protection of privacy of individuals vis a vis electronic data banks in the public sector referring both to "unfair discrimination".
133 Malgieri (n 132), 157 with reference to Damian Clifford and Jef Ausloos, 'Data Protection and the Role of Fairness' (2018) 37 *Yearbook of European Law* 130, 140 et seqq.
134 Malgieri (n 132), 157.
135 Rec. 39, 60, and 71 and Art. 13, 14, and 40 GDPR.
136 Cf. Rec. 60 GDPR stating "The principles of fair and transparent processing require that the data subject be informed of the existence of the processing operation and its purposes. The controller should provide the data subject with any further information necessary to ensure fair and transparent processing taking into account the specific circumstances and context in which the personal data are processed. Furthermore, the data subject should be informed of the existence of profiling and the consequences of such profiling. Where the personal data are collected from the data subject, the data subject should also be informed whether he or she is obliged to provide the personal data and of the consequences, where he or she does not provide such data. That information may be provided in combination with standardised icons in order to give in an easily visible, intelligible and clearly legible manner, a meaningful overview of the intended processing. Where the icons are presented electronically, they should be machine-readable." Cf. Jef Ausloos, Michael Veale and René Mahieu, 'Getting Data Subject Rights Right' (2019) 10(3) *JIPITEC* 283, 283.
137 Malgieri (n 132), 157 with reference to Smaranda Bara and Others v Casa Na ional de Asigur ri de S n tate and Others, Case C-201/14, [2015] (EU:C:2015:638); Opinion of Advocate General Campos Sánchez-Bordona delivered on 17 October 2018 (1); Deutsche Post AG v Hauptzollamt Köln, Case C-496/17, [2019] (ECLI:EU:C:2019:26).
138 Clifford and Ausloos (n 133), 140 et seq.
139 Information Commissioner's Office, 'Big data, artificial intelligence, machine learning and data protection Version 2.2' https://ico.org.uk/media/for-organisations/documents/2013559/big-data-ai-ml-and-data-protection.pdf (accessed 28 October 2020) 19 et seqq.; Michael Butterworth, 'The ICO and artificial intelligence: The role of fairness in the GDPR framework' (2018) 34(2) Computer Law & Security Review 257, 257 et seqq.; CNIL, 'Algorithms and artificial intelligence: CNIL's report on the ethical issues' (25 May 2018) https://www.cnil.fr/en/algorithms-and-artificial-intelligence-cnils-report-ethical-issues (accessed 28 October 2020).

includes also another—own—dimensions, *the mitigation of imbalances and prevention of discriminatory practices.*

Technical measures must be implemented that prevent data processing practices that might lead to discriminatory effects.[140] From a technical perspective the question remains what sort of technical measures are adept to discover discriminatory effects and mitigate them. The discovery and mitigation is a tricky if not impossible task because EU courts have interpreted and applied non-discrimination law heterogeneously.[141] It has therefore been claimed that the concept of fairness understood as non-discrimination cannot be implemented into automated systems: 'While numerous statistical metrics exist in the technical literature, it is not possible to reliably capture a European conceptualization of discrimination which is, by definition, contextual.'[142] This statement seems to focus in particular on cases of indirect discrimination where context matters most. In cases of direct discrimination (based on protected categories) non-context-related categories will be decisive.[143] While contextuality and flexibility of non-discrimination law and its interpretation is advantageous for many reasons (e.g., ensuring that the individual case receives the attention it deserves, that contextual factors such as time and relationships are reflected in the decision, that conflicting rights are balanced against each other, others), at the same time the contextuality of said laws renders their technical implementation impossible.[144]

These findings with respect to the technical implementation of fairness understood as the prevention of non-discriminatory practices lead to the conclusion that even if technical tools working towards fairness—in the use case for instance software that ensures the same accuracy rate of recognition of children faces irrespective of their ethnicity—can be employed, such tools will never fully be able to adhere to the fairness principle.[145] Taking the example of facial recognition, this is thus currently not possible, and it is likely that no system will ever be able to adhere to the principle.

3.5 Disentangling Connected Requirements

Requirements under the law are often connected across documents and domains. However, encoding them in a feasible and transparent way requires disentangling these dependency chains. One example thereof is the user-focused principle of transparency,[146] which includes two different elements, a prospective (incl. the continuous ability to have access to prospective information)[147] and a retrospective one.[148] While the former is an active information duty, the latter is more reactive and its scope has triggered a lively academic debate in particular on the establishment of a right to explanation[149] and the qualification of Article 22 of the GDPR[150] (see also Section 3.2 "'Solving' Conflicts in the Law").

The prospective information duty under the GDPR is *active*, meaning that the data controller must actively inform the data subject in an easily accessible manner (e.g., by way of a direct link, QR codes, SnapTags, NFC, dashboard) about the ongoing data processing. The burden of finding the information does not rest on the data subject.[151] To this end, the WP29 introduced the concept of push notice (i.e., just-in-time information notices) and pull notices (e.g., through a dashboard with the possibility to obtain further information).[152] From a design perspective it is key to avoid information overload,[153] which is why a layered approach to complying with the prospective information duty can be useful.[154] In itself, the prospective element contains multiple requirements which each trigger not only an individual implementation but one that puts each element into its bigger context.

One key information element is to *facilitate exercising individual rights* under Articles 15 to 22 of the GDPR.[155] Making use of one's individual rights does not require a specific motive; Curiosity about one's personal data being processed by a smart product must suffice to trigger an obligation of the data controller to provide said information.[156] A dashboard solution facilitates fulfilling this requirement and has been

140 This can be read into Rec. 71 GDPR explicitly mentions the use of technical measures to ensure that the processing does not lead to discriminatory effects.

141 Sandra Wachter, Brent Mittelstadt and Chris Russell, 'Why Fairness Cannot Be Automated: Bridging the Gap Between EU Non-Discrimination Law and AI' https://papers.ssrn.com/sol3/papers.cfm?abstract_id=3547922 (accessed 28 October 2020) 5 et seq.

142 Wachter, Mittelstadt and Russell (n 141), 5.

143 For further elaboration on the problem of antidiscrimination doctrine in the context of automated systems, see, e.g, Raphaële Xenidis and Linda Senden, 'EU Non-Discrimination Law in the Era of Artificial Intelligence: Mapping the Challenges of Algorithmic Discrimination' in Ulf Bernitz and others (eds), *General Principles of EU law and the EU Digital Order* (Kluwer Law International 2020), 151 https://ssrn.com/abstract=3529524 (accessed 28 March 2021); Frederik J. Zuiderveen Borgesius, 'Strengthening legal protection against discrimination by algorithms and artificial intelligence' (2020) 24(10) *The International Journal of Human Rights* 1572; Philipp Hacker, 'Teaching fairness to artificial intelligence: Existing and novel strategies against algorithmic discrimination under EU law' (2018) 55(4) *Common Market Law Review* 1143.

144 Wachter, Mittelstadt and Russell (n 141), 5 et seq.; cf. Hacker (n 143), 1146.

145 See Emre Kazim, Jeremy Barnett and Adriano Koshiyama, 'Automation and Fairness: Assessing the Automation of Fairness in Cases of Reasonable Pluralism and Considering the Blackbox of Human Judgment' https://ssrn.com/abstract=3698404 (accessed 28 March 2021).

146 Cécila de Terwangne, 'Article 5. Principles relating to processing of personal data', *The EU general data protection regulation (GDPR): A commentary* (OUP 2020) 314.

147 See here Article 29 Working Party, 'Guidelines on transparency under Regulation 2016/679' (WP 260 rev.01, 11 April 2018) at 10.

148 Frenzel (n 24), 21; Heike Felzmann and others, 'Transparency you can trust: Transparency requirements for artificial intelligence between legal norms and contextual concerns' (2019) 6(1) *Big Data & Society* 1, 2.

149 Cf. on the subject: Bryan Casey, Ashkon Farhangi and Roland Vogl, 'Rethinking Explainable Machines: The GDPR's 'Right to Explanation' Debate and the Rise of Algorithmic Audits in Enterprise' (2019) 34(1) *Berkeley Technology Law Journal* 143; Lilian Edwards and Michael Veale, 'Slave to the algorithm? Why a "right to an explanation" is probably not the remedy you are looking for' (2017) 16(1) *Duke Law and Technology Review* 18; Margot E Kaminski, 'The Right to Explanation, Explained' (2019) 34(1) *Berkeley Technology Law Journal*; Andrew D Selbst and Julia Powles, 'Meaningful information and the right to explanation' (2017) 7(4) *International Data Privacy Law* 233; Sandra Wachter, Brent Mittelstadt and Luciano Floridi, 'Why a right to explanation of automated decision-making does not exist in the general data protection regulation' (2017) 7(2) *International Data Privacy Law* 76.

150 Cf. e.g., Mendoza and Bygrave (n 99), 86 et seq.; Bygrave, 'Minding the Machine v.2.0: The EU General Data Protection Regulation and Automated Decision Making' (n 99), 246; Kaltheuner and Bietti (n 99), 10 et seq.; Martini (n 99), 29; Noto la Diega (n 99), 17.

151 Article 29 Working Party, 'Guidelines on transparency under Regulation 2016/679' (n 147), at 8.

152 Article 29 Working Party, 'Guidelines on transparency under Regulation 2016/679' (n 147), at 20 et seq.

153 Centre for Information Policy Leadership, 'Recommendations for Implementing Transparency, Consent and Legitimate Interests under the GDPR' (17 May 2017) https://www.informationpolicycentre.com/uploads/5/7/1/0/57104281/cipl_recommendations_on_transparency_consent_and_legitimate_interest_under_the_gdpr_-19_may_2017-c.pdf (accessed 28 October 2020) 2.

154 Ausloos, Veale and Mahieu (n 136), 286.

155 Art. 12(2) and 13(2)(b) GDPR.

156 Ausloos, Veale and Mahieu stating that individual rights are "intent-agnostic/motive-blind", Ausloos, Veale and Mahieu (n 136), 305 with reference to case law of national courts.

suggested by data protection authorities as well as scholars.[157] While a dashboard allows individuals to make use of their rights, such an action must trigger a predefined technical action in the background.[158] These actions will have to depend on the categories of data being processed. In fact, if a data subject consents only to a fully data-minimized processing (e.g., only locally stored data without third party or data controller access), making use of individual rights may become obsolete following Article 11 of the GDPR. From a design perspective, the exemption of Article 11 of the GDPR introduces a sort of hierarchy, as the provision indicates that the principle of data minimization must be given priority even if that means not being able to then fulfil individual rights. In many instances though, smart devices will rely on data processing of the data processor (e.g., use of external facial recognition software). Here, encoding data protection encounters technical constraints. In the concrete case of machine-learning-based facial recognition for instance, erasing the uploaded training data is possible, but erasing or rectifying inferences by machine-learning algorithms with respect to the classification is not feasible in general. Such 'unlearning' has become a topic of research in the machine-learning community,[159] however no satisfying approaches that can be applied generally exist to-date. In addition, similar to differential privacy systems, machine unlearning will imply trade-offs between the performance of a learning system and its unlearning ability.

3.6 Lack of Automatic Access to Relevant Structured Information

Prospective transparency duties extend to providing data subjects with information about what data is transferred to third countries and what adequacy measures are set in place to do so. Here, in a first step, one has to determine where (regarding geographical location) data flows when it 'leaves' a smart device. This becomes an issue when external processing is involved; The data sharing agreement should state where data is being processed in order to enable to determine automatically if the data is stored and processed in a country that falls under the 'adequacy decision list'[160] of the Commission. This list could also be automatically parsed by a computer system at regular intervals—in its current form with the help of heuristics that extract the individual country names from the list. It would, however, be desirable if regulatory information such as this was published in a machine-readable format and, ideally, would be linked to open data sources such as Wikidata that already contains representations of sovereign nations (e.g., representing the country Switzerland[161]). When data is not stored or processed in such an 'adequate' country or by a certified company, a device has to check whether binding corporate rules are in place that contain 'enforceable data subject rights and effective legal remedies for data subjects.'[162] These can take the form of standard contractual clauses adopted by the Commission,[163] which would be 'attached' to the data sharing agreements.[164] By means of Natural Language Processing (NLP) the agreements could be searched for such addendums and classified as such in order to provide a user with that information. Yet, this does not equal actual reading the agreements but merely provides for a fast way to check whenever data is processed in a country outside the adequacy decision list, if standard contractual agreements were signed. This would however require storing machine-processable representations of the contracts, which might often not be the case. One, albeit manual, possibility is to create and attach these documents in machine-readable formats (e.g., based on ODRL or LKIF, see Section 2.3 "Machine-understandable Data Protection Law") — this information could then be presented to users in a similar way to the transparency interface.[165]

While other approaches to fulfil this requirement exist, such as approved codes of conduct pursuant to Article 40 of the GDPR or certification mechanisms, the multitude of options complicates the technical codification of double checking whether this information requirement must be fulfilled and, if so, what information must be provided. Furthermore, to date, no standard format or mechanisms are established that could be used to implement automatic compliance checks of corporate rules or certificates and publication of which corporate rules or certificates that prove compliance with the GDPR. In other words, the four options to prove compliance if there are no adequacy decisions—namely binding corporate rules, use of standard contractual clauses, corporate rules,[166] or certifications—would require multiple additional steps and relying on information provided by the companies employing them and data protection authorities that are not easily available.

The challenges point also to policy-making needs: If encoding data protection in the spirit of Article 25 of the GDPR should become reality (or at least initiatives building towards it encouraged), measures that enable the extraction of relevant information is key. This requires an effort not only from data controllers, but also from data protection authorities to work towards standardizations and open-access of information that is published as machine-readable structured data

157 Cf. e.g., Article 29 Working Party, 'Guidelines on transparency under Regulation 2016/679' (n 147), at 10; cf. also Information Commissioner's Office, 'Guide to the General Data Protection Regulation (GDPR)' https://ico.org.uk/media/for-organisations/guide-to-data-protection/guide-to-the-general-data-protection-regulation-gdpr-1-0.pdf (accessed 28 October 2020) 90; cf. Philip Raschke and others, 'Designing a GDPR-Compliant and Usable Privacy Dashboard' in Marit Hansen and others (eds), *Privacy and identity management: The smart revolution: 12th IFIP WG 9.2, 9.5, 9.6/11.7, 11.4, 11.6/SIG 9.2.2 International Summer School, Ispra, Italy, September 4-8, 2017; revised selected papers* (IFIP Advances in Information and Communication Technology vol 526. Springer 2018).

158 Note that bystanders, whose image data is processed based on legitimate interests, do not have access to the dashboard needed to obtain information about the processing. To facilitate the information access and align with the principle of transparency, a visible QR code could be included onto the device's surface leading a bystander to further information about how data about non-users are being processed. This should take into account the concrete consent settings for the device in question which are stored by the data controller: thereby, bystanders would be informed about the concrete processing that their data undergoes.

159 Lucas Bourtoule and others, 'Machine Unlearning' (2020) https://arxiv.org/abs/1912.03817 (accessed 28 October 2020).

160 European Commission, 'Adequacy decisions' https://ec.europa.eu/info/law/law-topic/data-protection/international-dimension-data-protection/adequacy-decisions_en (accessed 28 October 2020).

161 https://www.wikidata.org/wiki/Q39 (accessed 28 October 2020).

162 Art. 46(1) GDPR.

163 Art. 46(1)(c) and (d) in conjuncture with Art. 93(2) GDPR.

164 Note that according to the ECJ's Schrems II decision the standard contractual clauses remain valid but it must be determined on a case by case basis whether in a particular transfer of data setting the clauses are legitimate. Cf. *Data Protection Commissioner v Facebook Ireland Ltd, Maximillian Schrems*, Case C-311/18, [2020] (ECLI:EU:C:2020:559), at para 134 and 149.

165 Bonatti and others (n 44), 1 et seqq.

166 In accordance with Art. 47 GDPR. Here, too, the verification that those are in place is not a straightforward issue that can easily be programmed. Technically, the simplest solution would be to verify whether the competent supervisory authority approved the corporate rules that have to fulfil a catalogue of requirements set out in Article 47(2) of the GDPR. However, this requires knowing which authority is in charge of approving the binding corporate rules of the external party and having said authority publish (and regularly update) a list elaborating which corporate rules it approved.

and thereby can directly be used by systems.

3.7 Dealing with Risk

The GDPR has intensified the debate on how to classify risks that occur with respect to the data protection rights of individuals.[167] On a macro-level two understandings must be differentiated: A broader interpretation of the risk-based approach applies the concept on both, compliance and enforcement of the GDPR; A more narrow understanding, applies it as an obligation targeted at data controllers.[168] On a more micro-level two further understandings of the risk-based approach must be differentiated: The WP29 approach separating between risks and compliance,[169] and Gellert's argument to understand risks as 'compliance risk.'[170] Even if only focusing on a micro-level, taking a risk-based approach requires differentiating between these two understandings. While the WP29 approach seems confusing and goal oriented (by acknowledging the need for flexibility as well as the danger of a risk-based approach for fundamental rights),[171] Gellert's approach relies upon the scalability notion of compliance.[172] He argues that two elements of risk must be differentiated: First, the event-element of a compliance risk is the lack of compliance altogether, and second, the consequence-element of compliance risk which is the resulting risk to the data subject's rights and freedoms.[173] On a meta-level, these interpretations show the challenges of dealing with risk when designing or even encoding data protection principles.

Even when dealing with the data security principle, where the measures that are specified in the law align with the technical understanding of how to keep data confidential, integer, and available at all times,[174] specifying the risks is not a trivial task. While the alignment of technical and legal objectives enables a more straightforward implementation of technical measures to achieve 'legal' aims, it remains difficult to automatically assess the internal and external risks and corresponding redress mechanisms. In fact, two steps are required to determine the engineering implications of the principle of data security. In a first step, the (external and internal) *risks of each data flow* including storage must be discussed. The risk will depend on the sensitivity of the data processed. For instance, biometric data (such as facial attributes) are more sensitive than other data. Therefore, the impact for the data subject if such data is exposed in a security incident is higher than for other data. In a second step, the *redress measures*, and the extent to which they minimize the outlined risk in the first step, must be described. Such measures include the erasure of training data after the training or only storing network credentials in an encrypted format and only for as long as they are required. With respect to the encryption format, future technological developments (also with respect to decryption schemes) must be taken into consideration.[175] This requirement leads to the responsibility to keep the system up to date—as mandated by Article 32 of the GDPR—which in turn implies a constant update of the recommended level of encryption according to established industry guidelines.

From a business perspective, a conscious weighing of strategic, user experience, and legal aspects (and risks) becomes necessary, which is hard to automate. It requires the data controller to balance the overhead in the design and implementation of the system, a possibly inferior user experience, and strategic business implications against the assumption of compliance risks and the overhead of properly managing collected information (e.g., secure storage, provisioning of data access to users, others). These decisions however need to be taken on a per-use-case, per-product, or even per-processing-purpose basis.

3.8 From Smartness to Dumbness?

An overly strict encoding of data protection principles – meaning that the necessity of much of the processed data is questioned and thus rejected – might lead to an overall reduction of the smartness of a device. In the extreme, this results in the design of a system that is unable to easily restore user passwords, thus undermining the positive perception of a product by users for the sake of maximizing the minimization of data collection. While such an extreme maximization of the data minimization principle goes against the inherent balancing notion of the GDPR, it is true that such an interpretation of the principle of data minimization and storage limitation can preclude several features of a product that are heralded as some of the prime advantages of 'digitalized business models.' For instance, if a device's location is not disclosed, the data controller cannot track its products (e.g., for supply-chain optimization). And if a device does not upload any diagnostics data, said data cannot be used by the data controller for product improvements or pre-emptive software updates or hardware repairs which might endanger the security of data and users; this also undermines rental and leasing business models. These modifications thus turn a 'smart device' into a more 'traditional' product. Moreover, an engineering decision to host the configuration dashboard locally instead of relying on a remote dashboard (i.e., a Website) to configure the system would lead to more complicated setups and higher cost on the side of the data controller while deteriorating the user experience. There are also *strategic* implications for the data controller: For example, the supplier of a smart device will need to weigh between its ambition to become independent of third-party facial recognition services (by storing uploaded images and using them to improve its own algorithms) and strict adherence to data minimization and storage limitation. The adherence to GDPR thus will require the data controller to find a balance between business aspects (e.g., ability to become independent of third-party services; ability to deliver an optimal user experience; ability to implement digital business models; others) and the legal risk and responsibility it assumes. This might, for some data controllers, lead to a 'minimal

167 Raphaël Gellert, 'Understanding the notion of risk in the General Data Protection Regulation' (2018) 34(2) *Computer Law & Security Review* 279, 279 et seq.; Lina Jasmontaite and others, 'Data Protection by Design and by Default: Framing Guiding Principles into Legal Obligations in the GDPR' (2018) (4)2 *European Data Protection L Rev* 168, 180 et seq.

168 Macenaite (n 21), 515.

169 Article 29 Working Party, 'Statement on the Role of a Risk-Based Approach in Data Protection Legal Frameworks' (WP 218, 30 May 2014), at 2.

170 Gellert (n 167), 284.

171 According to the WP29, individual data protection rights should be granted regardless of the level of risks of the processing and fundamental principles "should remain the same, whatever the processing and the risks for the data subjects." At the same time however, the WP29 also acknowledges that the fundamental principles are always applied in a context and are thus "inherently scalable." Moreover, the WP29 acknowledges that there are "different levels of accountability obligations depending on the risk posed by the processing in question." This statement is however again followed by a "but", as "controllers should always be accountable for compliance with data protection obligations including demonstrating compliance regarding any data processing whatever the nature, scope, context, purposes of the processing and the risks for data subjects are." Article 29 Working Party, 'Statement on the Role of a Risk-Based Approach in Data Protection Legal Frameworks' (n 169), at 3.

172 Gellert (n 167), 281 et seq.

173 Gellert (n 167), 282.

174 Tamò-Larrieux (n 32), 186 et seq.

175 Gerald Spindler and Philipp Schmechel, 'Personal Data and Encryption in the European General Data Protection Regulation' (2016) 7(2) *JIPITEC* 163, 172 with reference to Rec. 26 GDPR.

data design' where, in addition to the data that is absolutely required to leave the device for fulfilling its purpose (i.e., the images required for facial recognition), only information necessary for recording user consent is uploaded to the data controller, and might thus either undermine or in the extreme impede the data controller's business model.

In the extreme, data controllers could be motivated to only transiently process data in the hope that this qualifies as anonymous from the very beginning on (at the point of collection). The difficulties of achieving a state of full anonymization have been well documented, with various studies showing the identifiability of alleged anonymized data.[176] The GDPR though does not mandate full anonymity to fall outside its realm but a state of anonymization that is *not likely* to be reversed. To achieve this, one needs to not only look at the data itself (including the anonymized data), but also consider other resources that would reasonably enable re-identification.[177] This approach to anonymization under the GDPR has been criticized to overlook part of the risks of re-identification which are not only related to the data and resources available for identification but also depend on the motivation of the adversary to re-identify data, the security of the infrastructure in place, and the potential for mistakes that would lead to a disclosure allowing for re-identification.[178]

As mentioned, one measure that has been debated in the literature as a means to obtain anonymized data is *transient data processing*, i.e., technologies that merely sense their environment and process data ephemerally without storing it.[179] The legal reasoning that is key in this debate is the relative approach interpretation to personal data established by the ECJ.[180] In fact, a strict application of this approach would likely mean that transiently processed personal data that cannot be retrieved will fall outside the scope of the GDPR. Yet, following other interpretations of the term 'personal data,' such as the WP29 arguments that personal data can be established by purpose or result,[181] transient data processing may very well be covered under the GDPR.[182]

In any case, with respect to machine learning, transient processing can only relate to the raw data and not the learning aspect. Any transient data processed by machine learning algorithms influences the algorithms (the machine 'learned' something from it) and this derived or learned data (or parameters) is permanently kept within the system without the option to easily erase such derived data and undo its effects on the trained model.

Transient processing of the raw data can be combined with *local processing* such as image recognition with a pretrained local model as for instance Google's Inception-v3. If data can only be accessed by the owner of the device, it is questionable whether protection in this case is necessary. Similarly, the French Data Protection Authority (CNIL) argued that biometric data processing within smartphones falls under the *household exemption* if the biometric device is incorporated within a smartphone that only locally stores biometric templates of a user (e.g., fingerprints) and prevents the biometric data from being accessed from outside.[183] CNIL calls such a device an 'enclave' or 'sealed box.'[184] This reasoning does then not require an extensive analysis of whether personal data is being processed, but merely an analysis of whether data can be accessed from 'outside.' CNIL issued some rules for such technology to fall under the household exemption, such as: A user must use a device *privately*; the user has the *choice* to decide whether his or her data is being processed within the device (i.e., there must be alternative ways of unlocking a device in the case of biometric authentication); the data can by no means be *shared* with the outside (i.e., also external bodies cannot override this function); the stored data is *encrypted* by state of the art cryptographic algorithm and key management; and all technical solutions are *technically reliable*, i.e., the system is trustworthy.

These discussions show that encoding the principle of data minimization to its fullest can – depending on the design – result in avoidance of falling within the scope of the GDPR.[185] Yet, ephemeral processing of data also results in a reduction of the smartness of devices. How to balance these two aspects will depend on the context and purpose of processing. We see multiple examples where reduction of smartness

176 E.g., Latanya Sweeney, Akua Abu and Julia Winn, 'Identifying Participants in the Personal Genome Project by Name' (Data Privacy Lab, IQSS, Harvard University. White paper, 2013) https://privacytools.seas.harvard.edu/files/privacytools/files/1021-1.pdf (accessed 28 October 2020); Alexandra Wood, David O'Brien and Urs Gasser, 'Privacy and Open Data Research Briefing' https://papers.ssrn.com/sol3/papers.cfm?abstract_id=2842816 (accessed 28 October 2020); Article 29 Working Party, 'Opinion 05/2014 on Anonymisation Techniques' (WP 216, 10 April 2014).

177 With respect to anonymized data scholars have debated when encrypted data can be considered anonymous data. According to Spindler and Schmechel, encrypted data might only be anonymous data if only the data subject him or herself has access to the decryption key (but not in scenarios where the data controller still has access to both). The authors argue that in instances where the data controller does not have access to the decryption key, illegal attacks could still occur, yet that those do not have to be taken into account when determining if data is personal or anonymous. Cf. Spindler and Schmechel (n 175), 172 with reference to Opinion of Advocate General Campos Sánchez-Bordona, delivered on 12 May 2016, Case C-582/14 – *Patrick Breyer v Bundesrepublik Deutschland*. Cf. Rec. 26 GDPR.

178 Mark Elliot and others, 'Functional anonymisation: Personal data and the data environment' (2018) 34(2) *Computer Law & Security Review* 204, 205 et seqq. with further references.

179 Cf. Damian George, Kento Reutimann and Aurelia Tamò-Larrieux, 'GDPR bypass by design? Transient processing of data under the GDPR' (2019) *International Data Privacy Law* 285; Peter Davis, 'Facial Detection and Smart Billboards: Analysing the 'Identified' Criterion of Personal Data in the GDPR' (2020) *University of Oslo Faculty of Law Research Paper No. 2020-01* https://ssrn.com/abstract=3523109 (accessed 28 October 2020) 1; Maša Gali and Raphaël Gellert, 'Data protection law beyond identifiability? Atmospheric profiles, nudging and the Stratumseind Living Lab' (2021) 40 *Computer Law & Security Review* 105486.

180 *Breyer*, Case C-582/14, [2016] (ECLI:EU:C:2016:779). Note that the Breyer decision did not fully exclude the possibility of following an absolute approach. A vagueness that has been criticized by scholars.

181 Article 29 Working Party, 'Opinion 4/2007 on the concept of personal data' (WP 136, 20 June 2007), at 10; *Peter Nowak v Data Protection Commissioner*, Case C-434/16, [2017] (ECLI:EU:C:2017:994), at para 35 where the court argues that inferences about an individual are personal data as such information "by reason of its content, purpose or effect, is linked to a particular person."

182 Article 29 Working Party, 'Opinion 3/2012 on Developments in Biometric Technologies' (WP 193, 27 April 2012), 19 in which the WP29 states "it is not important to identify or verify the individual but to assign him/her automatically to a certain category." However, the WP29 does not mention whether such a categorization still involved the processing of personal data, "nor does it appear that the WP29 was cognisant of smart billboards that process data ephemerally"; cf. Davis (n 179), 11 et seqq.

183 CNIL, 'Biométrie dans les smartphones des particuliers: application du cadre de protection des données' (24 July 2018) https://www.cnil.fr/fr/biometrie-dans-les-smartphones-des-particuliers-application-du-cadre-de-protection-des-donnees (accessed 28 October 2020). We acknowledge that the ECJ has not decided on said issue and has traditionally taken a restrictive approach to interpreting the household exemption, cf. e.g., Urquhart and Chen (n 22) with further references. The ECJ has clearly stated that if data remains accessible to an unrestricted number of people or concerns public spaces this will not fall under the household exemption.

184 CNIL, 'Biométrie dans les smartphones des particuliers: application du cadre de protection des données' (n 183).

185 George, Reutimann and Tamò-Larrieux (n 179), 285 et seqq.

and even accuracy and traceability does not hinder achieving meaningful purposes (e.g., the COVID-19 tracking app based on D3PT,[186] or differential privacy models implemented by Google and Apple[187]). We believe that *leading by example* plays a crucial role in the field of legal code. It is however no surprise that DP3T and differential privacy models have emerged in academia. They require a time-consuming process and close collaboration between technical and legal researchers which are less likely to occur in companies that are driven by economic competition. The interdisciplinary collaboration though is central to these successes. The adoption of such technologies by states and companies shows their significant merit and demonstrates that 'imperfect remedies' might lead to good enough technology that balances different needs.

A path forward includes learning from these attempts to embed privacy protection into the design of technology and moves towards responsible technology by design. Achieving this requires a broader understanding and approach towards legal code and thinking about a *softer way of encoding legal principles* in a form that permits flexibility, transparency, and contestability.

4. Softcoding as a Path for More Responsiveness, Flexibility, and Transparency

As discussed in Section 2.3 "Machine-understandable Data Protection Law", the quest to encoding legal principles into software is not new and is currently gaining traction also outside of academia. From an industry standpoint, this would for instance enable more flexible variant management (e.g., when the same hardware is shipped to different legislations together with its firmware) and for facilitated adaptation of products to end users. The creation of machine-executable legal norms can also bring automation benefits to governments, for instance when aspects of regulation that include simple logic reasoning or mathematical operations are encoded. This is the case with New Zealand's Rates Rebate Act.[188] There, the government's Service Innovation Lab (LabPlus) wanted to rewrite the Rates Rebate Act (a tax rebate for low-income homeowners) in order to respond faster to citizen requests. To do so they first created pseudocode, which is still human-readable text but with defined consistent terminology. This pseudocode was then implemented as machine-executable instructions in the Python programming language. The LabPlus team stated in their final report that such an implementation is feasible for processes-oriented regulation (like the Rates Rebate Act) that involves 'factual information to determine application, eligibility, entitlement,' and prescribes a 'process that is used repeatedly' and one that 'can be delivered digitally.'[189] Similar initiatives can be found world-wide, with for example the OECD issuing a recent working paper on 'Rules as Code' which likewise promotes the creation of machine-consumable law.[190] In addition, researchers have even started to experiment with machine-learning systems that attempt to forecast decisions in case law.[191] How promising those attempts are, remains to be seen.[192]

The examples show that even if legal scholars lament the imperfectness of the interlinking of code and law these instruments are being created and deployed. In that sense, it is not a matter of 'whether' design-based regulation should be employed, but much more on '*how*' we want it to be developed.

While the issues in Section 3 "Encoding Data Protection: An Imperfect Remedy" are mostly of translational nature, they point to two further clusters of challenges: *System-related* and *moral* ones. Addressing the challenges also means taking into account the different ways legal code can be implemented.

4.1 How Softcoding Mitigates Some of the Translational Challenges

The *translational challenges* show that law is more than just written text. It is constantly interpreted, adjusted to a specific context, and adapts over time. However, this is not true for all legal provisions either: The law is not vague in every aspect. Moreover and from a data controller's perspective, regulators could – if a need arises – be more precise, and even publish aspects of regulation (e.g., encryption standards, tax rebates calculations, or lists of countries that are considered safe to transfer data to) in a machine-readable way so that this information can be readily consumed by software and acted upon. What that means is that translational issues should be resolved by taking steps towards the middle ground and asking what norms can and cannot - and should and should not - be made more amenable.

While softcode does not help per se to deal with translational issues (e.g., how to ensure that no generalizations are projected into the code, no assumptions are made on how to interpret the law, etc.), it allows for systems to be more transparent, malleable, and responsive. Such decoupling thus enables a system to adapt over time to its regulatory environment; enabling change is an important aspect to deal with translational issues, in particular in light of how interpretations of law may change over time. The system's higher responsiveness that derives from the decoupling of major decision parameters through softcoding would thus simplify the updating of the system and thereby reduce the probability that the system remains non-compliant.

4.2 How Softcoding can Address System-Related and Moral Challenges

On a broader perspective, *system-related challenges* arise with respect to *who* should be in charge of developing code that adapts to its legal environment and how transparent such code is made to the public. The New Zealand example shows clearly a collaboration effort and involvement of the government to achieve a machine-executable Rates Rebate Act. Other initiatives, like the one the authors are

186 Cf. https://github.com/DP-3T/documents (accessed 28 October 2020).
187 Cf. https://developers.googleblog.com/2019/09/enabling-developers-and-organizations.html and https://www.apple.com/privacy/docs/Differential_Privacy_Overview.pdf (accessed 28 October 2020).
188 https://www.digital.govt.nz/dmsdocument/95-better-rules-for-government-discovery-report/html (accessed 8 November 2020).
189 Service Innovation Lab (LabPlus), 'Better Rules for Government, Discovery Report' (March 2018) 27 https://www.digital.govt.nz/dms-document/95-better-rules-for-government-discovery-report (accessed 8 November 2020).
190 OECD Working Papers on Public Governance, 'Cracking the code: Rulemaking for humans and machines' (2020) available at https://www.oecd-ilibrary.org/governance/cracking-the-code_3afe6ba5-en (accessed 20 December 2020).
191 Cf. Kevin D Ashley, 'A Brief History of the Changing Roles of Case Prediction in AI and Law' (2019) 36(1) *Law in Context* 93, 103 et seqq.
192 E.g., in Estonia the idea of implementing AI judges was raised. However, no official information on the success or failure of this project can be found. A news article on said topic dates back to 2019: Eric Niller, 'Can AI Be a Fair Judge in Court? Estonia Thinks So' (*Wired*, 25 March 2019) https://www.wired.com/story/can-ai-be-fair-judge-court-estonia-thinks-so/ (accessed 11 November 2020). Another example is the CaseCruncher Alpha, an artificial intelligence that became famous through a challenge where it was able to predict the outcome of cases with greater accuracy than the lawyers involved: Rory Cellan-Jones, 'The robot lawyers are here - and they're winning' (*BBC News*, 1 November 2017) https://www.bbc.com/news/technology-41829534 (accessed 8 November 2020).

following in an implementation of data compliant code is based on open-source software and decoupled, standardized legal vocabularies and ontologies and can thus in principle be held under scrutiny by users and judges alike. Yet, companies will likely not promote open-source legal code initiatives. As Herbert Burkert said already in 1997 with respect to privacy-enhancing technologies (PETs): 'PET design must be open to participatory elements. This implies designing PETs and implementing them in social systems must involve those whom these enhancements are supposed to serve.'[193]

Adopting a softcoding approach, for instance by coupling code with openly accessible ontologies that render regulation machine-readable, the data controller opens the possibility to let the end user fine-tune compliance settings of smart products, thereby increasing transparency and participation. It is even conceivable that individual agents in the society create and publish carefully crafted alternative legal ontologies that subclass a legal domain's legislation and might go beyond it (or might selectively ignore aspects of it to enable disobedience, see below). Like-minded individuals could then further develop and share these documents and point their own smart products towards them.

In addition, softcoding could help to address *moral challenges* that arise predominantly because of the lack of engagement or choice of an individual when confronted with techo-regulation. Mireille Hildebrandt talks here about a lack of buffer between the rules and the one who is ruled; in her own words: 'Rather, under the Rule of Law the legal system acts as a buffer between ruler and ruled, creating the possibility to contest state-authority in an appeal to a court that is in fact supported by the authority of the state (the paradox of the Rechtsstaat).'[194] The crucial functionality represented by a buffer is the preservation of the option of (civil) disobedience.

The ability to disobey is fundamental to moral agency. Moral agency requires the freedom to act and vulnerability with respect to the consequences one suffers if one breaks the rule.[195] Freedom to act can be impaired by legal code; yet does not have to. Karen Yeung describes three scenarios using the same road safety technology: Code that automatically stops a car at red lights. The scenarios then differ by the goals three individuals are trying to pursue: A *criminal minded-person*, who wants to cross a road at red to hurt others; a *person* who masters self-restraining most of the time but sometimes does cross at a red light; and a *Good Samaritan* who wants to cross at red to help someone else in an emergency situation. Yeung shows that the criminally-minded person still has agency to harm others in other ways; that the self-restraining person loses physical agency but not moral one (even though that person will not get praise for abiding the law without the legal code); and that the Good Samaritan has to determine other means to achieve her or his goal, but can still be seen as morally praiseworthy independent on the action he or she chooses (i.e., other means or riding the car to the hospital despite the red lights).

The discussion shows that the moral challenges should not be described with broad brushstrokes. To the contrary, they require a nuanced discussion. Softcoding approaches must be open and flexible enough to preserve the possibility for disobedience. Design-based regulation should thus not lead to 'regulation by technology'[196] but what we could call '*regulation nudged by technology*.' Such as speeding cameras that nudge individuals to comply with the speed limit while driving, technology can by default nudge individuals to comply with the rule yet allow for informed disobedience as well as contestability of those parameters (e.g., speeding to ensure that a woman in labor gets to the hospital in due time and contesting the rule due to an emergency situation). While the default value can be compliance, it must be made easy to modify the technology to - in certain instances - not comply with the rule.

In contrast to hard-coded legislation, softcoding approaches that couple a system with a default legal ontology that can be replaced by the user preserve the ability of the individual, and of society, to exert civil disobedience. The Good Samaritan from Yeung's example would be able to point her car at an ontology that does not regiment it into stopping at a red light, or one where this behavior can be overridden by the user. In principle, she could also create such a version of the machine-readable regulation herself or together with others, and publish it. Such folksonomy-based approaches would thereby pave the way to keep society in the loop.[197]

4.3 Calling for Transdisciplinary Experts

Lastly, while the literature to encoding data protection principles has proposed both, bottom-up and top-down approaches,[198] we believe that bottom-up approaches, which require legal, implementation, and business strategy teams to engage in interdisciplinary communication and collaboration are more fruitful and enable meta-deliberation processes that are much needed in the field of legal (soft) code. In contrast to top-down approaches, iterative and bottom-up approaches encourage a deeper cross-disciplinary understanding and creative solution finding. This aligns more with the reality that open-text legal documents bring along such as ambiguity that leaves room for case-by-case interpretation by legal professionals who need to interpret facts of a case given subjective words or phrases and in the context of national and international legislation that might be connected to the investigated text corpus through opening clauses. While interdisciplinary collaboration is the starting point, we believe that there is a need to train transdisciplinary experts that that can 'deal with emerging value conflicts'[199] arising from the deployment of new technologies. Such transdisciplinary experts should be equipped with tools and strategies to resolve value conflicts and promote the design of responsible technology.

5. Conclusion

Neither hardcoding nor softcoding of regulation into software systems and cyber-physical systems are perfect. In contrast to hardcoding, where regulation is hard-wired into code at a given time and cannot be easily adjusted when regulation changes, softcode attempts to tie code to regulation through loose coupling. This can be accomplished for instance by means of ontologies that are publicly accessible and interpretable by users. Yet, no matter whether regulation is soft- or hardcoded, various issues remain: The need to encode

193 Herbert Burkert, 'Privacy-Enhancing Technologies: Typology, Critique, Vision' in Philip E. Agre and Marc Rotenberg (eds), *Technology and Privacy: The New Landscape* (MIT Press 1997) 125, 135.

194 Hildebrandt, 'Legal Protection by Design: Objections and Refutations' (n 7), 236.

195 Yeung (n 2), 9 et seqq.

196 Cf. Hildebrandt, 'Legal Protection by Design: Objections and Refutations' (n 7), 247 et seq.

197 Cf. on the idea and implementation of society-in-the-loop Iyad Rahwan, 'Society-in-the-loop: Programming the algorithmic social contract' (2018) 20 *Ethics and Information Technology* 5.

198 Cf. Section 2.3 "Machine-understandable Data Protection Law"; cf. e.g., Jaap-Henk Hoepman, 'Privacy Design Strategies' in *29th IFIP International Information Security Conference* (Marrakech, Morocco, June 2014); Seda Gürses, Carmela Troncoso and Claudia Diaz, 'Engineering Privacy by Design Reloaded' in *Amsterdam Privacy Conference* (Amsterdam, The Netherlands, 2015).

199 Lutz and Tamò (n 70).

assumptions because of a lack of clarity in the law, to resolve conflicts within legal norms, and to generalize terms in order to ensure compliance remain critical problems that arise. The advantage of softcode with respect to those issues is only that the system can be improved and changed over time to adapt to new legal circumstances (e.g., court decisions that have clarified legal terms and solved specific conflicts).

These issues are of translational nature, but go beyond the pure translation of law into code as they trigger systemic and moral issues as well. Systemic issues arise from a lack of transparency and the actors involved in the creation of legal code. While here, too, softcode provides some remedies, depending on how legal code is created (based on deterministic or more probabilistic decision-making systems) and by whom (state-driven initiatives vs. industry-driven ones), the opacity of legal code will remain. However, softcode would open the possibility of creating transparency tools that would enable developers and also laypersons to inspect the legal code that drives their products. Furthermore, moral issues are triggered by the lack of engagement between the ruler and the one who is ruled. Crucially, this lack of engagement can curtail civil disobedience which is key to allow social change within a society. With softcode, and the civil disobedience that it can guarantee on the individual and societal levels through folksonomy-enabled meta-disobedience, these moral issues can in principle be overcome.

Overall, the findings within this article point thus to the need for a broader yet more nuanced discussion. Future research needs to map and investigate the current designed-based regulation deployment and initiatives, their effect on individuals and society at large, their openness, the architectural decoupling of implementations and legal code, the involved decision-making (deterministic vs. probabilistic approaches), and the actors involved in the design of legal code. To do so requires not only expertise in computer science and law but calls upon the expertise of multiple disciplines within the social science community.

Acknowledgements

This work was supported by the Hasler Foundation (project #19086). The authors thank Damian George for his helpful comments on earlier versions of this article as well as Emre Bayamlıoglu and Ronald Leenes for their insightful comments during the Digital Legal Talks.

03

Technology and Regulation

On the legal responsibility of artificially intelligent agents: addressing three misconceptions

Antonia Waltermann*

(legal) responsibility, moral agency, legal agency, liability, artificially intelligent agents.

antonia.waltermann@maastrichtuniversity.nl

This paper tackles three misconceptions regarding discussions of the legal responsibility of artificially intelligent entities: (a) that they cannot be held legally responsible for their actions, because they do not have the prerequisite characteristics to be 'real agents'; (b) they should not be held legally responsible for their actions, because they do not have the prerequisite characteristics to be 'real agents'; (c) they should not be held legally responsible for their actions, because to do so would allow other agents to 'hide' behind the AI and thus escape responsibility. (a) is a misconception not only because (positive) law is a social construct, but also because there is no such thing as 'real' agency. The latter is also the reason why (b) is misconceived. The arguments against misconceptions a and b imply that legal responsibility can be constructed in different ways, including those that hold both artificially intelligent and other (human or corporate) agents responsible (c). The paper concludes that there is more flexibility in the construction of responsibility of artificially intelligent entities than is at times assumed.

1. Introduction

The emergence and proliferation of artificially intelligent entities (hereafter referred to also as artificial agents or AI) raises questions of legal liability and responsibility. This is because some artificially intelligent entities do not require human input to perform some action, nor do their actions necessarily follow pre-programmed patterns. Given the developments in machine learning, it seems that (some) artificial agents are acting autonomously and that more artificial agents will be acting more and more autonomously in the future.[1] This leads to an accountability gap in the law.[2] Situations in which harm occurs for which no one is responsible according to current positive law (*lex lata*) but which, it seems, should not have to be borne by the entity suffering it are becoming increasingly likely. How this accountability gap should be closed has been subject to much debate, both politically and academically.[3] In this paper, I will focus on three (interconnected) misconceptions within these debates.[4] Most references will be to tort law, but the ground for legal responsibility, be it tort, contractual, or criminal, does not matter. The three misconceptions are that artificially intelligent entities:

A. *cannot* be held legally responsible for their actions, because they do not have the prerequisite characteristics to be 'real agents' and therefore cannot 'really' act.

B. *should not* be held legally responsible for their actions, because they do not have the prerequisite characteristics to be 'real agents' and therefore cannot 'really' act.

C. *should not* be held legally responsible for their actions, because to do so would allow other (human or corporate) agents to 'hide' behind the AI and escape responsibility that way, while they are the ones who should be held responsible.

The first two misconceptions are connected by the content of the argument put forward ("AI lack the prerequisites to be 'real agents'") but differ in the kind of conclusion that is justified by it, the first conceptual and the second normative. Meanwhile, the second and third misconception are connected by the conclusion of the argument ('AI should not be held legally responsible') but differ with regard to the content of the argument put forward to justify that conclusion.

This paper argues that all three arguments (a-c) are misconceived. The argument to this effect proceeds along the following lines: first, I will briefly outline what I mean by artificially intelligent entities (section 2). Then, I will elaborate on the first misconception (a) that

1 By this, I mean that they act in ways that are not foreseen or predicted and not (easily) foreseeable or predictable. At times, this may go hand in hand with not being (easily) understandable or explainable by programmers/developers. Some more on this in section 2.

2 Cf. Gunther Teubner, 'Digitale Rechtssubjekte? Zum Privatrechtlichen Status Autonomer Softwareagenten' (2018) *Archiv für die civilistische Praxis;* Susanne Beck, 'The Problem of Ascribing Legal Responsibility in the Case of Robotics' (2015) 31 *Ai & Society* 473.

3 Cf. European Parliament resolution of 16 February 2017 with recommendations to the Commission on Civil Law Rules on Robotics (2015/2103(INL)); 'Open Letter to the European Commission Artificial Intelligence and Robotics' http://www.robotics-openletter.eu accessed 26/01/2021; Francisco Andrade and others, 'Contracting Agents: Legal Personality and Representation' (2007) 15 *Artif Intell Law* 357; Joanna J. Bryson, Mihailis E. Diamantis and Thomas D. Grant, 'Of, for, and by the People: The Legal Lacuna of Synthetic Persons' (2017) 25 *Artif Intell Law* 273 for an overview of some political proposals, calls, and concerns.

4 This focus mirrors Ugo Pagallo, 'Apples, oranges, robots: four misunderstandings in today's debate on the legal status of AI systems' (2018) *Phil. Trans. R. Soc.*, although the misconceptions addressed and the arguments used to address them differ.

* Antonia Waltermann is assistant professor at Maastricht University, the Netherlands. I am very grateful for useful comments by Jaap Hage, Daniel On, and Rūta Liepina, as well as by the anonymous reviewers. The usual disclaimers about the final result apply.

Received 9 Apr 2021, Accepted 3 Jul 2021, Published 12 Jul 2021.

legal agency must (conceptually) coincide with 'real' agency (section 3). This is a misconception not only because (positive) law is a social construct, but also because there is no such thing as 'real' agency (section 4). The latter is also the reason why the second argument (b) is misconceived. The argument that there is no 'real' agency will require an excursion into the realm of philosophy and the cognitive sciences,[5] but as I hope to demonstrate, this excursion is highly relevant to the question whether legal responsibility of AI is possible and desirable.

The arguments against misconceptions a and b imply that legal responsibility can be constructed in different ways, including those that hold *both* artificially intelligent and other (human or corporate) agents responsible (section 5), pre-empting the concern that human/corporate agents could 'hide' behind AI responsibility (misconception c). Accordingly, this paper concludes that there is more flexibility in the construction of responsibility of artificially intelligent entities than is at times assumed (section 6). This offers more freedom to law- and policymakers, but also requires openness, creativity, and a clear normative vision of the aims they want to achieve.

Before diving into the argument of the paper, some caveats and clarifications are required.

This paper deals with questions of responsibility and agency, but these terms are used in different contexts with different meanings. In computer science, for example, an agent is an entity that "observes the world through sensors and acts upon an environment using actuators" and "directs its activity toward achieving goals in a rational manner" or, in more technical terms, [a]n agent is a system that receives at time t an observation O_t and outputs an action A_t."[6] Law, meanwhile, knows the concept of an agent in agency law, where a person (the agent) acts as representative of another person (the principal), for example when a lawyer negotiates a contract on behalf of a client. In philosophy of action and in ethical theory, agent again means something else (see section 4). Where this paper uses the term 'agent', this is never in the sense of agency law; instead, the focus is on agents as entities capable of acting (in a sense relevant for responsibility).

When it comes to the terms 'liability' and 'responsibility', a common distinction is between legal liability on the one and moral responsibility on the other hand. Departing from this, I will use '(legal) responsibility' throughout this paper as an umbrella term for all types of liability. Similarly, I will use 'responsible' instead of 'liable'. Even where I omit the prefix 'legal' of 'legal responsibility', I will refer to legal responsibility, as opposed to moral responsibility, unless otherwise stated. In many areas of law (e.g. contract and tort), it would be more accurate to speak of liability than responsibility, but in other areas (e.g. international law), the term responsibility is used. I consider responsibility the more suitable term for the purposes of this paper to indicate a. the proximity to questions of moral responsibility and b. the abstraction from a particular legal field.

The latter relates to a point I want to further emphasise: the argument of this paper is situated at a high level of abstraction: it is not an argument about any particular legal system[7] or area of law. Instead, it is an argument about the relation between law, legal concepts, and concepts and insights from the cognitive sciences broadly construed.

2. Artificially intelligent entities

The European Commission defines artificial intelligence as follows:

> 'Artificial intelligence (AI) refers to systems that display intelligent behaviour by analysing their environment and taking actions – with some degree of autonomy – to achieve specific goals.
>
> AI-based systems can be purely software-based, acting in the virtual world (e.g. voice assistants, image analysis software, search engines, speech and face recognition systems) or AI can be embedded in hardware devices (e.g. advanced robots, autonomous cars, drones or Internet of Things applications).'[8]

For the purposes of this paper, whether an artificial agent is purely software-based or physically embedded is not relevant; both purely software-based agents such as algorithms used in, for example, insurance risk assessment, and physically embedded ones such as autonomous vehicles or weapons systems can cause harm of the kind that raises questions of (legal) responsibility.

A distinction often made in this connection concerns different levels of autonomy (or independent action) of the artificially intelligent entity: 'from human supervision (level 1), and deterministic autonomy (level 2), to machine-learning (level 3) and multi-agent systems (level 4).' An alternative distinction that focuses on the level of human involvement is between human in the loop, human on the loop (equivalent to level 1) and human out of the loop (ranging from levels 2 to 4). In cases of 'human in the loop', human input is required before an action can be performed. In cases of 'human on the loop', actions can and will be performed without human input, but there is human supervision, and the supervising human can override the artificial agent's decision before the action is performed. An example of this would be a self-driving car with a human 'supervisor' who can redirect the car, or a weapon system that requires authorisation from a human being. In cases of 'human out of the loop', finally, there is no human input or interaction. Here, distinctions can be made between those cases where there is prior human input and the algorithm performs the 'loop' according to deterministic programming (level 2), to those scenarios where the algorithm is capable of learning and adapting its behaviour to what it has learned in ways not anticipated by programmers/designers. One could think of an autonomous car that learns to model its behaviour from other road users, for example. If an autonomous car also communicates with other autonomous cars and adapts its behaviour to information – such as road conditions or the location and length traffic jams – from other autonomous cars, this would be an example of a multi-agent system.[9]

The degree of autonomy is relevant when it comes to the accountability gap in law: current legal instruments, concepts, and arrangements do not seem sufficient for increasingly autonomous artificial agents.

5 I use cognitive sciences here in a very broad sense, including - but not limited to - neuroscience, psychology, and behavioural economics.

6 Woodrow Barfield, 'Towards a law of artificial intelligence' in Woodrow Barfield and Ugo Pagallo (eds), *Research Handbook on the Law of Artificial Intelligence* (Edward Elgar Publishing 2018); Daniel Silver, Satinder Singh, Doina Precup, Richard S. Sutton, 'Reward is enough' (2021) *Artificial Intelligence* 299, 3.

7 Although the author's background is in civil rather than common law, which will be reflected in some of the examples chosen. Nonetheless, the questions raised and argument made (should) hold mutatis mutandis.

8 Communication from the Commission to the European Parliament, the European Council, the Council, the European Economic and Social Committee and the Committee of the Regions on Artificial Intelligence for Europe, Brussels, 25.4.2018 COM(2018) 237 final.

9 Antje von Ungern-Sternberg, 'Artificial Agents and General Principles of Law' (Available at SSRN: https://ssrn.com/abstract=3111881) *German Yearbook of International Law*, 4 f.

This is because the potential solutions that can currently be found in positive law often require a certain level of control and foreseeability by the human or corporate agent producing, owning, or using the artificially intelligent entity or require that the human or corporate agent has acted in a wrongful or culpable way before holding that (corporate or human) agent legally responsible. In cases of contractual breach, for example, an autonomous software agent cannot be held liable according to current German law, given that the software agent lacks the legal capacity to act (rechtliche Handlungsfähigkeit). Consequently, if the (human or corporate) operator of the software agent can demonstrate that they did not themselves violate a contractual obligation, there is no liability, and the other contracting party is left with the damage of the contractual breach caused by the software agent. A similar gap exists with regard to tort liability.[10] More generally, Barfield summarises that 'the use of artificial intelligence begs the question of who is liable if the artificial intelligence controlling smart technology learns and solves problems in ways completely unknown to the human in the system' and '[t]he more autonomous the system, that is, the more the human is removed from the decision-making loops of the system, the more difficult for courts to assign liability to humans when there is a system failure.'[11]

The above gives a broad definition of artificially intelligent entities and outline of the problem, but for the argument of this paper, nothing more specific is required.

3. Misconception a: legal agency must (conceptually) coincide with 'real agency'

The first misconception I tackle in this paper can be summarised as follows: AI cannot be held legally responsible because AI is not an agent.

Coeckelbergh, for example, indicates that

> 'a problem that becomes especially relevant in the case of AI is attribution of responsibility. Since technologies cannot be responsible moral agents and are hence a-responsible, the only way to ensure responsible action is to make humans responsible.'[12]

Dahiyat writes that

> 'Some commentators think that software agents are merely coded information and that we will commit excessive conceptual mistakes if we attribute a legal or moral responsibility to these agents, or if we just assume that they possess whatever else we take to be present when we hold human beings responsible for their actions. This is because, unlike humans who are sensitive, self-determined and moral, software agents lack a number of conditions, which should be fulfilled in order for responsibility to be ascribed.'[13]

Statements such as these indicate, it seems to me, that legal agency must (conceptually) coincide with 'real' agency.[14] In response to this, I will argue that (positive) law as a social construct is (conceptually) independent from any perceived 'real' agency, that is, that law can technically regard entities as legal agents even if they are not 'real' agents. The mere technical possibility, however, does not mean that the law should do so. This is addressed in section 4.

Brozek and Jakubiec identify a spectrum of possible positions regarding the legal responsibility of artificially intelligent entities. The two extremes of this spectrum are 'restrictivism' and 'permissivism'. Restrictivism 'denies the possibility of holding autonomous machines legally responsible on purely metaphysical grounds'[15] while permissivism 'imposes no restrictions on the possible legal constructions'[16]. Restrictivism[17] denies the possibility for holding artificially intelligent entities legally responsible on the grounds that they lack essential qualities necessary for legal (and moral) responsibility.[18] Candidates for these essential qualities are consciousness, intentionality and the capacity for intentional action, (libertarian) free will, autonomy, the capacity for deliberation, alignment between one's reasons for action (in the sense of justificatory reasons, not heuristics or causes) and one's actions, and more. In more legal terminology, AI cannot be held responsible because it lacks both Handlungs- and Schuldfähigkeit, that is, the capacity to act and be culpable.[19]

The restrictivist argument[20] indicates that

(P1) An entity lacking *xyz* characteristics cannot be legally responsible.

(P2) Artificially intelligent entities lack *xyz* characteristics.

(C) Artificially intelligent entities cannot be legally responsible

This presumes that certain entities, possessing certain characteristics, are 'real' agents and 'really' responsible and that the law must conceptually coincide with this extra- or pre-legal reality, that is, that law must accurately map this external[21] reality.

This notion that law (and its concepts) must coincide with extra-legal reality and that it is not (technically) possible for law to do otherwise is clearly a misconception. This is supported by the view that (positive) law is a social construct,[22] which makes it technically possible

10 Teubner (n 2); Gunther Teubner, 'Rights of Non-Humans? Electronic Agents and Animals as New Actors in Politics and Law' (2007) 04 *Max Weber Lecture Series*; Beck (n 2).
11 Woodrow Barfield, (n 6). The chapter offers a number of concrete examples of challenges to the current legal situation.
12 Mark Coeckelbergh, 'Artificial Intelligence: Some Ethical Issues and Regulatory Challenges' (2019) *Technology and Regulation*, 31, cf. Mark Coeckelbergh, 'Artificial Intelligence, Responsibility Attribution, and a Relational Justification of Explainability' (2020) 26 *Science and Engineering Ethics* 2051.
13 Emad Abdel Rahim Dahiyat, 'Law and Software Agents: Are They "Agents" by the Way?' (2020) *Artif Intell Law*, 67.
14 I do not here want to attribute this exact misconception to any particular author. I do, however, want to suggest that it is implicit in the argumentation of many. If I am mistaken about this, all the better.
15 Bartosz Brozek and Marek Jakubiec, 'On the Legal Responsibility of Autonomous Machines' (2017) 25 *Artif Intell Law* 293, 294.
16 Ibid.
17 I use restrictivism and restrivists throughout the following sections and attribute certain views to restrictivists/restrictivism. This should not be taken as a claim that all authors that hold some restrictivist views necessarily hold all the views I here describe. As Brozek and Jakubiec (n 15) point out, this is one extreme on a spectrum of possible views and approaches. An uncharitable interpretation of my approach is that I am constructing and arguing against a strawman, but even if no one were to hold a strictly restrictivist view, it is useful to consider the misconceptions this view rests on. Using the extreme for this purpose serves to highlight the misconceptions.
18 Brozek and Jakubiec (n 15), 294.
19 There is variation in terminology and concepts between different legal fields here; I hope readers will forgive the generalisation.
20 This is essentially what Solum calls the "missing-something" argument applied to legal responsibility, rather than personhood: Lawrence B Solum, 'Legal Personhood for Artificial Intelligences' (1992) *North Carolina Law Review* 70 (4).
21 External to the law, in this case.
22 This sentence does not contain a commitment to a positivist concept of law, as non-positivist law theories account for positive law as a social construct as well. Hage, for example, convincingly argues this point in Jaap Hage, 'The Limited Function of Hermeneutics in Law' in David Duarte, Pedro Moniz Lopes and Jorge Silva Sampaio (eds), *Legal Interpreta-*

to give it any content whatsoever. Brozek and Jakubiec describe it as 'quite possible from [a] purely technical point of view, since the law is a conventional tool of regulating social interactions and as such can accommodate various legislative constructs, including legal responsibility of autonomous artificial agents'.[23] Many others have made the same point in a variety of contexts, not limited to the legal responsibility of artificially intelligent entities.[24] Moreover, differences between different legal systems and cultures as well as across time further support this point: here, one can think of criminal responsibility of animals in the Middle Ages,[25] the legal positions of slaves e.g. in times of the Roman Empire or of the legal position of women in Western societies until quite recently.[26] Lastly, another example is the personhood of anything, 'be it monasteries or corporations, governments or ships in maritime law, rivers in New Zealand or India, down to the entire ecosystem in Ecuador.'[27]

This response to the restrictivist claim that legal concepts must coincide with extra-legal reality leaves open the possibility that there are 'real' agents that can 'really' be responsible and other entities that cannot 'really' be responsible because they lack the essential characteristics for 'real' responsibility. All this response posits is that it is *technically possible* to regard an entity as a legal agent, irrespective of whether it is a 'real' agent or not. *Legal* agency is a legal construct.

This leaves room for a counterargument from the restrictivist perspective: while it may be technically possible for the law to construct legal agency any way it wants, it *should not* do so. Instead, the law should only regard those entities as agents that are 'real' agents, and it should only hold those entities responsible that are 'really' responsible. In other words: law should model its constructs after 'real' agents. Generally, this argument proceeds along the following lines: there are a number of characteristics such as intentionality, autonomy, consciousness, or free will, that are required for 'real' agency and responsibility.[28] Because artificially intelligent entities lack these capacities, they cannot 'really' be responsible agents; instead, human beings can and should be held morally and legally responsible –because they meet these conditions and are 'really' responsible agents.[29]

This is the second misconception I will tackle.

4. Misconception b: legal agency should coincide with 'real agency'

The second misconception, that the law should not attribute responsibility to artificially intelligent entities because these entities are not or cannot be 'real' agents or 'really' responsible rests on the assumption, as pointed out above, that there is such a thing as a 'real' agent or 'real' responsibility.[30]

Intuitively, the idea that there are real agents that are responsible for their actions and that we human beings are such responsible agents makes sense: we distinguish between agents – those entities that make things happen and go through the world seemingly independently of physical laws – and non-agents, things like rocks and puddles or other inanimate objects that behave in predictable ways and are clearly and obviously subject to natural laws.[31] We perceive other human beings as agents whose actions are more accurately and more easily explained by their desires and intentions than by physical laws acting upon them. Not only that, but we also perceive ourselves as agents causally responsible for our actions which are shaped not by physical laws acting upon us, but by our desires and intentions – and we often perceive our actions as something we have willed, something that was the result of our wanting and deciding to do something.[32] Moreover, we are responsible for our intentional and free actions. As Solum already indicated in his seminal paper on legal

tion and Scientific Knowledge (Springer 2019) 5.
Of course, a non-positivist might argue that while it is technically possible for positive law to have any content whatsoever, positive law may well be *wrong*. Depending on the specific non-positivist theory, this may go hand in hand with the claim that the positive law is then not law at all, meaning that it is not, in fact, possible for *law* to have any content whatsoever. While section 4 of this paper does not use non-positivist language, I think it can be taken to address this claim with minor (mental) translations by the non-positivist reader.

23 Brozek and Jakubiec (n 15), 303.

24 For example Hans Kelsen, *General Theory of Law and State* (Harvard University Press 1945), 94 and Bartosz Brozek, 'The Troublesome Person' in Visa Kurki and Tomasz Pietrzykowski (eds), *Legal Personhood: Animals, Artificial Intelligence and the Unborn* (Springer 2017), 8 with regard to natural persons, see also Ngaire Naffine, 'Who Are Law's Persons? From Cheshire Cats to Responsible Subjects' (2003) 66 *The Modern Law Review* 346; Ulfrid Neumann, 'Strafrechtliche Verantwortlichkeit Von Verbänden – Rechtstheoretische Prolegomena' in Klaus Volk, Klaus Lüderssen and Eberhard Kempf (eds), *Unternehmensstrafrecht* (De Gruyter 2012), 16 with regard to corporate criminal responsibility. More generally, cf. Alf Ross, 'Tû-Tû' (1957) 70 *Harvard Law Review* 812.

25 Piers Beirnes, 'The Law Is an Ass: Reading E.P. Evans' the Medieval Prosecution and Capital Punishment of Animals' (1994) 2 *Society and Animals* 27; William Ewald, 'Comparative Jurisprudence (I): What Was It Like to Try a Rat?' (1995) 143 *University of Pennsylvania Law Review* 1889.

26 Married women in the Netherlands, for example, could not perform legal acts without the consent of their husbands until 1957. This example is taken from Robert van den Hoven van Genderen, 'Legal Personhood in the Age of Artificially Intelligent Robots' in Woodrow Barfield and Ugo Pagallo (eds), *Research Handbook on the Law of Artificial Intelligence* (Edward Elgar Publishing 2018).

27 Ugo Pagallo, 'Vital, Sophia, and Co.—the Quest for the Legal Personhood of Robots' (2018) 9 *Information* 230, 9. In my view, arguing analogously from personhood to agency is possible (but not vice-versa) because personhood (generally) presumes agency (but not vice-versa).

28 Dorna Behdadi and Christian Munthe, 'A Normative Approach to Artificial Moral Agency' (2020) 30 *Minds and Machines* 195. While there is debate on whether agency presupposes responsibility and distinctions are made between conditions for (moral) agency and (moral) responsibility, I will not consider these questions here and instead talk about 'responsible agents'. Himma, for example, argues that under the standard view (which I turn to in this section), consciousness is a condition for responsibility, but that 'the very notion of agency itself presupposes consciousness in the sense that only a conscious being can be an agent', Kenneth Einar Himma, 'Artificial Agency, Consciousness, and the Criteria for Moral Agency: What Properties Must an Artificial Agent Have to Be a Moral Agent?' (2009) 11 *Ethics and Information Technology* 19, 28 and Coeckelbergh, 'Artificial Intelligence, Responsibility Attribution, and a Relational Justification of Explainability' (n 12) holds (for human beings) that 'agency is normally connected with responsibility. You have an effect on the world and on others, and therefore you are responsible for what you do and for what you decide.'

29 Coeckelbergh, 'Artificial Intelligence, Responsibility Attribution, and a Relational Justification of Explainability' (n 12), 2055.

30 This assumption can be found e.g. in Bryson et al (n 3) with regard to legal personhood. Gunkel outlines how under one view, blaming artificially intelligent entities is 'ontologically incorrect', David J. Gunkel, *The Machine Question: Critical Perspectives on Ai, Robots, and Ethics* (MIT Press 2012) 28. Dahiyat (n 13) holds that 'we will commit excessive conceptual mistakes if we attribute a legal or moral responsibility to these agents'; Coeckelbergh, 'Ethics of artificial intelligence: Some ethical issues and regulatory challenges' (n 12) holds that 'only humans can be responsible agents'.

31 Samir Chopra and Laurence White, *A Legal Theory for Autonomous Artificial Agents* (University of Michigan Press 2011) 11; Joshua Greene and Jonathan Cohen, 'For the Law, Neuroscience Changes Nothing and Everything' (2004) 359 *Philosophical Transactions: Biological Sciences* 1775, 1782.

32 Patrick Haggard and Valerian Chambon, 'Sense of Agency' (2012) 22 *Curr Biol R* 390; Patrick Haggard and Manos Tsakiris, 'The Experience of Agency' (2009) 18 *Current Directions in Psychological Science* 4.

personhood for artificial intelligence, '[o]ur understanding of what it means for a human being to function competently has ties to our views on responsibility'.[33] Fischer and Ravizza describe our ordinary concept of moral responsibility as follows:

> 'An important difference between persons and other creatures is that only persons can be morally responsible for what they do. [...] Whereas both persons and non-persons can be causally responsible for an event, only persons can be morally responsible. [...] [I]n order to be praiseworthy or blameworthy a person must know (or be reasonably expected to know) what he is doing, and he must not be deceived or ignorant about the circumstances and manner in which he is doing it. [...] A second type of excusing condition is force. [...] [A]n agent has the type of freedom necessary to be morally responsible only if he has 'control over his actions,' the act is 'up to him,' he was 'free to do otherwise,' he 'could have acted differently', and so forth.'[34]

They also indicate that 'there seems to be a difference between being *held* responsible and actually *being* responsible.'[35] While it may be possible to *hold* artificially intelligent entities legally responsible, one could say, they *are not* actually responsible – and therefore should not be *held* to be.[36]

The understanding of ourselves as responsible agents I have sketched above takes our (subjective) experience and intuitions as central. As such, it can be termed 'phenomenological'. Phenomenology 'address[es] the meaning things have in our experience, [...] as these things arise and are experienced in our 'life-world'.'[37] This intuitive understanding of ourselves as responsible agents is reflected also in philosophy of action and the notion of moral agency in normative ethics, fields that seek to theorise, systematise, and critically reflect on the intuitions that we have and our social and normative practices.[38] Philosophy of action does so with regard to when an event is an action and when an entity is an agent, normative ethics with regard to when an action is right, wrong, good, bad, permissible, or impermissible or, more generally, with the moral evaluation of actions.[39] The standard understanding of action here holds that human beings are (the only) real agents[40] and that something is an action if it is intentional:[41]

> '[i]ntuitively, an agent is something able to take actions. One way to distinguish agents from other entities is that agents do things, as opposed to have things happen to them; to deny something or someone agency is to deny the capacity to take actions, for the actions of the agent distinguish it from the rest of the world. [...] Related to this notion is the concept of self-directed actions or acting for reasons, for the philosophical sense of 'agency' is linked with the ascription of intentions. To possess agency is to be the originator of action, to be driven by motivations, purposes, desires, and autonomously, freely-chosen decisions.'[42]

According to the standard view, 'moral agents must meet rationality, free will or autonomy, and phenomenal consciousness conditions'.[43] Human beings are 'real' agents because we are capable of acting intentionally, freely, and autonomously, and we are 'really' responsible for our intentional and free actions,[44] that is, because we (seemingly) fulfil these conditions. One aspect of this view is what can be termed (naïve) realism about agents and responsibility: the idea that there are 'real' agents irrespective of (moral or legal) agency-ascriptions and that there is such a thing as 'real' responsibility that is different from being *held* responsible on the basis of moral, social, or legal norms.

33 Solum (n 20).

34 John Martin Fischer and Mark Ravizza, 'Introduction' in John Martin Fischer and Mark Ravizza (eds), *Perspectives on Moral Responsibility* (Cornell University Press 1993) 4. Himma (n 28) identifies this as the standard view: 'for all X, X is a moral agent if and only if X is (1) an agent having the capacities for (2) making free choices, (3) deliberating about what one ought to do, and (4) understanding and applying moral rules correctly in paradigm cases.'

35 Fischer and Ravizza (n 34), 18.

36 This is reflected, for example, in Dahiyat (n 13) and Coeckelbergh, 'Ethics of artificial intelligence: Some ethical issues and regulatory challenges' (n 12). See Behdadi and Munthe (n 28) for an overview of this approach when it comes to moral responsibility.

37 David Woodruff Smith, 'Phenomenology' in Edward N. Zalta, *The Stanford Encyclopedia of Philosophy* (Summer 2018), https://plato.stanford.edu/archives/sum2018/entries/phenomenology, 1.

38 Consider e.g. Fischer and Ravizza (n 34), 7: 'A theory of moral responsibility ought to accommodate these standard excusing conditions in the sense that the ascriptions of responsibility entailed by the theory ought to match our ordinary intuitions about when agents are and are not morally responsible.' John Hyman, *Action, Knowledge, and Will* (Oxford University Press 2015), 32 argues that these fields go (even) further than our intuitive understanding in a kind of 'chauvinism' about action.

39 Julia Driver, *Ethics: The Fundamentals* (Blackwell Publishing 2007), 2.

40 Hyman (n 38), 30. Of course, law holds non-human entities such as corporations responsible. This may be permissible under this view because these composite entities are then, in a sense, 'derived' agents: they derive their agency and responsibility from the fact that one or more human beings have acted. This implies that regarding human beings as legal agents and holding them legally responsible rests on their 'real' agency, while regarding composite entities such as corporations or states as legal agents and holding them legally responsible rests on a legal fiction. Conceiving of corporations and states as such 'derived' agents is, under this view, permissible because they are composed of human beings, the paradigmatic, 'real' agents. For artificially intelligent entities, however, this is not the case. In particular in 'human out of the loop'-scenarios, there is no human agent from whom to derive agency and responsibility. Brozek and Jakubiec (n 15) for example, make this point. Cf. also Jiahong Chen and Paul Burgess, 'The Boundaries of Legal Personhood: How Spontaneous Intelligence Can Problematise Differences between Humans, Artificial Intelligence, Companies and Animals' (2019) 27 *Artif Intell Law* 73 regarding spontaneous intelligence.

41 More specifically, that something is an action if it is intentional under some description or if it is identical to or derived from an intentional action. Markus Schlosser, 'Agency' in Edward N. Zalta, *The Stanford Encyclopedia of Philosophy* (Fall 2015), https://plato.stanford.edu/archives/fall2015/entries/agency, para 2. What this means is that if you unknowingly alert a burglar by intentionally turning on the light, alerting the burglar is an action of yours because it is either the same action as turning on the light under a different description (after all, you alerted the burglar by turning on the light) or it is derived from your intentional action of turning on the light. For the purpose of this paper, not much rides on whether an event is an action if it is intentional under some description or identical to or derived from an intentional action; what matters is that intentional action is the fundamental conception of action on this view. Not all philosophers of action take this view. Hyman (n 38), for example, argues that intentionality is not decisive.

42 Chopra and White (n 31), 11 f.

43 Behdadi and Munthe (n 28), 197.

44 This is a broad outline that does not leave room for nuanced differentiation between different theories. For a more elaborate overview on agency, see e.g. Schlosser (n 41) or Matt King and Peter Carruthers, 'Responsibility and Consciousness' in Derk Pereboom and D.K. Nelkin (eds), *Oxford Handbook on Moral Responsibility* (Oxford University Press, forthcoming). An overview of different views related to actions and responsibility can be found in Joseph Keim Campbell, Michael O'Rourke and Harry S. Silverstein, *Action, Ethics, and Responsibility* (Bradford Books 2010) and Fischer and Ravizza (n 34). The standard view of (moral) agency is often contrasted to the functionalist view, under which 'agency requires only particular behaviors and reactions which advocates of the standard view would view as mere indicators of the capacities stressed by the standard view.' Behdadi and Munthe (n 28), 197. I focus here on the standard view, as that is the view underlying the misconception I am addressing.

Given this, the second misconception can be rephrased as follows: *'really' responsible agents exist and law should only ascribe agency and responsibility to those entities that are 'real' agents.* Is it likely, however, that we are 'real' agents' and 'really' responsible in the way our intuitions and the standard view indicate? And are our intuitions and the phenomenological view sufficient basis for making choices about the (legal) ascription of agency and responsibility?

I argue that they are not. My argument rests on insights from the cognitive sciences broadly construed that suggest that the phenomenological view is misguided, particularly as concerns (naïve) realism about agency and responsibility. This implies that our intuitions about ourselves and the criteria for responsible agency are not as strong a justification for choices about the legal ascription of agency and responsibility as we assume. In the following, I briefly touch on a number of different arguments that challenge the distinct 'realness' of human agency.[45]

There is increasing evidence that there are two systems for human decision-making, including moral and legal decision-making: one that is unconscious, fast, and instinctive or automatic, the other conscious, slower, and controlled.[46]

> 'Dual-process theories of thinking and reasoning quite literally propose the presence of two minds in one brain. The stream of consciousness that broadly corresponds to System 2 thinking is massively supplemented by a whole set of autonomous subsystems in System 1 that post only their final products into consciousness and compete directly for control of our inferences, decisions and actions.'[47]

That we sometimes make 'gut decisions' and sometimes carefully consider our choices may not seem particularly radical or challenging to the (phenomenological) view we have of ourselves as agents. What is challenging is the degree to which we make choices unconsciously and to which biases and heuristics play a role in those choices we think we have made rationally and without any other factors at play, according to dual-process theory and the evidence substantiating it. Implicit biases such as racism or sexism have a large impact on our judgments and behaviour, as in this 2005 study:

> 'Subjects were asked to rate the suitability of two candidates for police chief, one male and one female, where one candidate was presented as 'streetwise' but lacking in formal education while the other one had the opposite profile. Despite the fact that Uhlmann and Cohen varied the sex of the candidates across conditions – so that some subjects got a male streetwise candidate and a female well-educated candidate while other subjects got the reverse – subjects considered the male candidate significantly better qualified in *both* conditions. [...] Rather than being conscious of the sexist attitude, the agent is conscious of a confabulated criterion which itself seems plausible – i.e. the importance of being streetwise or highly educated.'[48]

Beyond that, situational factors shape our behaviour in ways we are not aware of, such as a scramble-sentence test including words relating to rudeness makes subjects considerably more likely to interrupt a conversation (67%) than the control group (38%) or those subjects whose scramble-sentence test included words related to politeness (16%); the presence of a briefcase (as opposed to a backpack) triggering more competitive behaviour; or the time since the last food break having significant impact on how judges ruled in decisions relating to prison parole.[49]

Neuroscientific studies have corroborated the dual-process theory and found neurobiological correlates.[50] These insights challenge the presupposition that we are generally rational and that all, most, or even many of our actions are intentional. Further evidence that our intuitions about our own actions and their causes are far less reliable than they seem to us comes from insights related to confabulation. Carruthers indicates that '[t]here is extensive and long-standing evidence from cognitive and social psychology that people will (falsely) confabulate attributions of judgments and decisions to themselves in a wide range of circumstances.'[51] This evidence indicates that we are 'inaccurate in reporting the *causes* of [our] judgments or behavio[u]r' and decisions. For instance, subjects of an experiment instructed to move a finger and to freely decide which finger upon hearing a noise reported that they had decided to move the finger that they moved – but the actual cause of the digit moving was focal magnetic stimulation of areas of the relevant motor cortex areas. These subjects believe that they have acted on the basis of an intentionally made choice, that is, that they are the ('real') agent, but this is a confabulation.[52] Our intuitions about our actions being intentional are not reliable. Specifically with regard to our sense of agency (defined as the experience of controlling one's own actions and thereby events in the world), Haggard and Chambon write that this experience of agency can be tricked and is sometimes illusory.[53]

The assumption that our intention is causally relevant for our actions, that is, that our intentional choices cause, direct, and guide our actions, is further called into question by insights from and following from the Libet experiments. In these experiments, it was found that a 'readiness potential' for action in the brain preceded not only the voluntary movement, but also awareness of the conscious intention to move.[54] Some of these results have been interpreted in such a way that consciousness plays less or even no causal role when it comes to our actions. This is also the conclusion of the social psychologist Daniel Wegner who holds that

> 'each human mind has an abbreviated view of itself, a self-portrait

45 These arguments will necessarily brief and behind each of them is a discussion that cannot be reproduced here in full. My aim here is not to give an exhaustive account of the insights, debates, and nuances; to do so would go far beyond the scope of this paper. The arguments mainly refer to empirical, rather than philosophical insights, although I agree with authors such as Caruso that 'philosophical arguments on their own are sufficient for showing that people are never morally responsible for their actions in the basic desert sense' (Gregg Caruso, 'If Consciousness Is Necessary for Moral Responsibility, Then People Are Less Responsible Than We Think' (2015) 22 *Journal of Consciousness Studies*, 54). I will not reiterate these arguments here.

46 Cf. Daniel Kahneman, *Thinking, Fast and Slow* (Farrar, Straus and Giroux 2011); Joshua David Greene, Moral Tribes: Emotion, Reason, and the Gap between Us and Them (Penguin Press 2013); Jonathan St B. T. Evans, 'In Two Minds: Dual-Process Accounts of Reasoning' (2003) 7 *Trends in Cognitive Sciences* 454.

47 Evans (n 46), 458.

48 Caruso (n 45), 52.

49 Caruso (n 45), 54.

50 Evans (n 46), 455.

51 Peter Carruthers, 'How We Know Our Own Minds: The Relationship between Mindreading and Metacognition' (2009) 32 *Behavioral and Brain Sciences* 121, 130.

52 Ibid, 131, reviewing, inter alia, Nisbett and Wilson (1977), Brasil-Neto et al. (1992) and Wegner & Wheatley (1999).

53 Haggard and Chambon (n 32).

54 Benjamin Libet and others, 'Time of Conscious Intention to Act in Relation to Onset of Cerebral Activity (Readiness-Potential): The Unconscious Initiation of a Freely Voluntary Act' (1983) 106 *Brain* 623.

that captures how it *thinks* it operates, and that therefore has been remarkably influential. The mind's self-portrait has as a central feature the idea that thoughts cause actions, and that the self is thus an origin of the body's actions. This self-portrait is reached through a process of inference of *apparent mental causation*, and it gives rise to the experience that we are consciously willing what we do. Evidence from several sources suggests that this self-portrait may often be a humble and misleading caricature of the mind's operation—but one that underlies the feeling of authorship and the acceptance of responsibility for action.'[55]

These interpretations are debated, particularly when making the strong claim that consciousness plays no causal role whatsoever; nonetheless, they offer further support for the thesis that we are far less intentional and conscious agents than we think and that while we have a feeling of authorship and responsibility, such feelings do not offer privileged information about causal responsibility. Another element of the phenomenological view as outlined above is that unlike rocks, stones, or even more complex ordinary matter such as bees or mice, we have the power to freely bring about one event or some alternative event, that is, the power to do otherwise.[56] This understanding of freedom to choose between events is in conflict with causal determinism and quantum indeterminacy, thereby further calling into question the phenomenological view.[57]

While none of these arguments and insights by themselves prove that the phenomenological view and its notions of 'real' agency and 'real' responsibility are mistaken, they demonstrate that the presuppositions of this view and the intuitions that support it are neither as plausible nor as solid as our (unexamined) intuitions may make them appear. Given this, our intuition that some entities (namely human beings) are 'real' agents which can be 'really' responsible does not provide a good argument against ascribing legal agency and responsibility to other entities (that is, AI) by itself: the insight that our intuitions and our understanding of ourselves – of what causes our actions and decisions – are often based on mistaken confabulations calls into question the phenomenological view and thereby also the normative implications that should attach to it. If we are often wrong about our understanding of ourselves and others as responsible agents, if there are good reasons to doubt the accuracy of our intuitions, why should we attach normative consequences solely to the belief that we are 'real' agents and other entities are not?

To be clear, I am not making an argument that we should disregard our intuitions entirely. I am making the argument that it does not suffice to say 'law should not attribute agency and responsibility to AI *because AI are not 'real' agents or 'really' responsible*'. Instead, it seems to me that a normative argument that does not rely solely on our – likely mistaken – intuitions and reference to the phenomenological view is required.[58] What could such an argument look like? One example can be found in Brozek and Jakubiec who argue that while it is possible for law to attribute agency and responsibility to AI, it should not do so because this would take law too far from the life-world and therefore, any such rules would remain 'law in book' rather than 'law in action'.[59] This is an argument from legal efficacy and our intuitions and phenomenological view. Whether it is the case that any such rules would be inefficacious is an as of yet unanswered empirical question.[60] This argument demonstrates, however, that to call into question the phenomenological view's presuppositions does not necessarily mean negating or disregarding the fact that people do have the intuitions that feature in the phenomenological view. Instead, the demand for an argument that goes beyond the phenomenological view indicates a different place for these intuitions in the argument: they are empirical information that needs to be embedded in a normative argument, instead of indicators of absolute, external truth.

Another example of a normative argument of the kind I have in mind as necessary in the debate whether law should attribute agency and responsibility to AI is the following:

> '[A]scribing responsibility to software agents might hide the real source of the problem, mask the human creator of the harm, and might also be used as an excuse for some people to evade their responsibility and behave recklessly.'[61]

This argument, found more frequently in the literature,[62] can be rephrased as follows: law should not attribute agency and responsibility to artificially intelligent entities because to do so would allow other (human or corporate) entities to escape responsibility in cases in which they (the human or corporate entities) should be held responsible.

This brings us to the third misconception I want to address in this paper.

5. Misconception c: hiding behind AI responsibility

There are (at least) two possible ways to demonstrate that it is a misconception to believe that holding AI responsible would necessarily allow other agents to escape responsibility: this can be demonstrated by looking at (conceptual) possibility and by looking at current legal practice.

The first approach to the second misconception relates back to the point made in section 3. of this paper: (positive) law is socially con-

55 Daniel M. Wegner, 'The Mind's Self-Portrait' (2003) 1001 *Annals of the New York Academy of Sciences* 212. Wegner holds further that '[e] xperiences of conscious will thus arise from processes whereby the mind interprets itself – not from processes whereby mind creates action. Conscious will, in this view, is an indication that we think we have caused an action, not a revelation of the causal sequence by which the action was produced.' Summary taken from Daniel M. Wegner, 'Frequently Asked Questions About Conscious Will' (2004) 27 *Behavioral and Brain Sciences* 679; see also Daniel M. Wegner, 'The Mind's Best Trick: How We Experience Conscious Will' (2003) 7 *Trends in Cognitive Sciences* 65; Daniel M. Wegner, *The Illusion of Conscious Will* (MIT 2002).

56 Fischer and Ravizza (n 34), 8.

57 Fischer and Ravizza (n 34) offer an overview of this incompatibility as well as the different positions that have been taken in the debate, mainly libertarianism and compatibilism. See also "Jaap Hage and Antonia Waltermann, 'Responsibility, Liability, and Retribution' in Bartosz Brozek, Jaap Hage and Nicole Vincent (eds.), Law and Mind: A Survey of Law and the Cognitive Sciences (Cambridge University Press 2021).

58 The call for a normative approach when it comes to the (in this case moral) responsibility of artificially intelligent entities can be found also in Behdadi and Munthe (n 28). The arguments leading to the conclusion of their article and mine strike me as compatible and can be read in conjunction.

59 Brozek and Jakubiec (n 15), 293.

60 My intuition on this question is a different one than that of Brozek and Jakubiec: I believe such rules could very well be(come) efficacious, in part because it seems to me that we take the intentional stance quickly, in part because law influences our life-world. Cf. S. Marchesi and others, 'Do We Adopt the Intentional Stance toward Humanoid Robots?' (2019) 10 *Front Psychol* 450.

61 Dahiyat (n 13), 69.

62 Bryson et al (n 3) consider it the main case of potential abuse and (rightly) point out that lawmakers must provide solutions for this. See also Gunkel (n 30).

structed. Its rules are created (be it by legislators such as parliaments, or by judges), which means that we (read: our law creators) can set up the system in such a way that it works for us,[63] as well as change it if it has adverse effects or does not lead to the desired results.[64]

Accordingly, it is – technically, in theory – possible to attribute agency and responsibility to more than one entity. Whether this is desirable and for what reasons it is or is not desirable cannot be addressed in this paper but understanding the ontological nature of agency and responsibility (both within and outside of the law) as a social construct allows us to understand the degree of control that we (or in this case: our lawmakers) have over the situation.

In how far is it necessary to adapt existing laws and legal concepts to do so?

When it comes to tort liability, the law already knows circumstances in which more than one entity is regarded as the tortfeasor. Landes and Posner distinguish between 'simultaneous' and 'successive' joint tort: the first covering those cases where 'the victim suffers a single or indivisible injury as a result of the tortious activity of two or more parties',[65] and the second covering those cases where 'one tortfeasor aggravates an injury inflicted by the other, as where a driver negligently hits a pedestrian and a physician negligently treats, thereby aggravating, the pedestrian's injury'.[66] In the Principles of European Tort Law,[67] Title V outlines rules for multiple tortfeasors, either under solidary or under several liability:[68]

Art 9:101 Solidary and several liability: relation between victim and multiple tortfeasors

1) Liability is solidary where the whole or a distinct part of the damage suffered by the victim is attributable to two or more persons. Liability is solidary where:

   a) a person knowingly participates in or instigates or encourages wrongdoing by others which causes damage to the victim; or

   b) one person's independent behaviour or activity causes damage to the victim and the same damage is also attributable to another person.

   c) a person is responsible for damage caused by an auxiliary in circumstances where the auxiliary is also liable.

2) Where persons are subject to solidary liability, the victim may claim full compensation from any one or more of them, provided that the victim may not recover more than the full amount of the damage suffered by him.

3) Damage is the same damage for the purposes of paragraph (1)(b) above when there is no reasonable basis for attributing only part of it to each of a number of persons liable to the victim. For this purpose it is for the person asserting that the damage is not the same to show that it is not. Where there is such a basis, liability is several, that is to say, each person is liable to the victim only for the part of the damage attributable to him.

These already existing conceptual tools could, it seems to me, be employed to prevent a situation in which corporate or human agents escape liability, although outlining the specific form this should take goes beyond the scope of this paper.[69] When it comes to criminal liability, it is similarly true that more than one person can be liable as principal, with notions such as joint perpetration, perpetration-by-proxy, instigation, and aiding further delineating situations of multiple agents.[70] However, in criminal law, matters are made more complicated by the fact that some legal systems construe the act requirement for criminal liability more stringently and at times less explicitly normatively than when it comes to tort or other liability, such as Germany regarding corporate criminal liability, for example.[71] This is a subject for another paper and cannot here be addressed. Equally, tort liability and criminal liability are not the only liability regimes that one could and should consider when it comes to responsibility of artificially intelligent entities.[72] For present purposes, however, it suffices to say that there are means, both when it comes to *lex lata* and *lex ferenda*, to ensure that attributing legal responsibility to artificially intelligent agents does not allow other agents, human or corporate, to escape responsibility.

This demonstrates that it is not *necessarily* true that AI responsibility would preclude the responsibility of other agents. Whether AI should be held responsible and the most suitable means of implementing such responsibility in practice if it is found to be desirable are important matters for both academic and political discussion, but not the aim of this paper. In this paper, I only seek to address a limited number of misconceptions, not give all-things-considered recommendations or conclusions.

6. Conclusion

This paper has addressed three misconceptions regarding the legal agency and responsibility of artificially intelligent entities: first, that law cannot attribute agency and responsibility to such entities because they are not 'real' agents or 'really' responsible; second, that it should not do so for the same reason; third, that if the law were to attribute agency and responsibility to such entities, it would allow other (human or corporate) agents to escape responsibility, while

63 It is more complicated than that, of course: what rules will have what impact is at times very difficult to predict. Moreover, parliaments are not single entities but composed of different individuals belonging to different political parties, which may pursue different aims. And so on. Nonetheless, the general point stands.

64 Beck (n 2) offers different possibilities, including some discussion of advantages and disadvantages.

65 William M. Landes and Richard A. Posner, 'Joint and Multiple Tortfeasors: An Economic Analysis' (1980) 9 *The Journal of Legal Studies* 517, 518. This can be further divided into 'joint care' and 'alternative care' cases, that is, cases in which both parties have to take care to avoid the damage occurring, and cases in which it would be sufficient if only one party had taken care.

66 Ibid, 518.

67 While the Principles of European Tort Law are non-binding guidelines, they try to merge different traditional approaches with a modern perspective on how the law of torts should develop in the future and as such provide a good exemplification of what concepts of tort law exist and may be implemented in the future in Europe.

68 For a comparative law overview of multiple tortfeasor liability in Europe, W. V. H. Rogers and W. H. van Boom, *Unification of Tort Law: Multiple Tortfeasors* (Kluwer Law International 2004).

69 Cf. Lewis A Kornhauser and Richard L Revesz, 'Sharing Damages among Multiple Tortfeasors' (1989) 98 *The Yale Law Journal* for a law and economics approach to different liability regimes and their potential effects in situations involving multiple tortfeasors.

70 Cf. Laura Peters, *Acting Together in Crime* (Eleven International Publishing 2018).

71 The German view is that corporations can neither act nor be culpable and that they lack the capacity for both. Therefore, Germany does not know corporate criminal liability. Instead, an administrative (quasi-criminal) approach is used. David Roef (2019) 'Corporate Criminal Liability' in Johannes Keiler and David Roef (eds) *Comparative Concepts of Criminal Law* (Intersentia 2019).

72 The possible contractual liability of artificial agents should not be disregarded, for example; the possible legal responsibility of autonomous weapons in humanitarian law situates questions of agency- and responsibility-ascription (also) in the international legal sphere.

they should be held responsible.

Given that (positive) law is a social construct, it is clearly technically possible for law to attribute agency and responsibility to artificially intelligent entities. Legal historical and comparative legal research shows that this has been done; legal theory demonstrates why it can be done. However, the mere technical possibility does not mean it should be done. The second misconception argues that agency and responsibility should be attributed to 'real' responsible agents, presupposing that there are such 'real' and 'really' responsible agents. This presupposition, I have argued, fits with the phenomenological view of the world and our place in it, as well as the standard view on agency and responsibility: we (human beings) are the paradigmatical responsible agents because we possess consciousness, intentionality, and rationality. However, insights from the cognitive sciences demonstrate that the presuppositions of this view and the intuitions that support it are neither as plausible nor as solid as we may assume. Given this, I have raised the question why we should attach normative consequences to the belief that we are 'real' agents and other entities are not in itself? The view that our intuitions about 'real' agency are not in themselves sufficient basis for refusing to attribute agency and responsibility to artificially intelligent entities does not necessitate disregarding these intuitions; they can inform normative arguments and be embedded in them.

A normative argument against attributing legal agency and responsibility to artificially intelligent entities is that it would allow other agents (human or corporate) to hide behind the artificially intelligent entities and escape responsibility that way, while they should be held responsible. However, understanding that (legal) agency and responsibility are constructed also means that who is regarded as an agent in law and held responsible can be changed in such a way as to produce the desired consequences. This includes the possibility to hold both artificially intelligent agents *and* human and/or corporate agents responsible *at the same time*. Investigating whether this should be done and if so, what form this should take goes beyond the scope of this paper, but there is no technical or conceptual impossibility to do so.

Artificially intelligent entities pose a challenge for policy- and lawmakers due to the accountability gap they create. This paper has addressed three misconceptions in debates about one possible means to close the accountability gap, namely, to regard artificially intelligent entities as agents responsible for their own acts. As such, the explicit scope of this paper has been relatively narrow. Nonetheless, I think that implicitly, this paper also demonstrates another challenge that artificially intelligent entities pose (for policy- and lawmakers, scholars, citizens, and so on): by investigating how (legal) concepts do (or do not) apply to artificially intelligent entities, we have to address our assumptions about ourselves and our place in the world, especially where these are not as accurate as we have long thought. This requires intellectual humility[73] but at the same time, understanding the ontological nature of (legal) agency and responsibility, both that of artificially intelligent entities and ourselves, as a social construct allows us to understand the degree of control that we (or in this case: our lawmakers) have over the situation. It shows us the freedom we have to shape and create practices of agency and responsibility that suit our (normative) goals. Thus, there is more flexibility in the construction of responsibility of artificially intelligent entities than one might assume, which offers freedom to law- and policymakers, but also requires openness and creativity as well as a clear, normative vision of the aims we and they want to achieve.

73 Cf. Kathryn Schaffer and Gabriela Barreto Lemos, 'Obliterating Thingness: An Introduction to the "What" and the "So What" of Quantum Physics' Foundations of Science' (2019) *Foundations of Science*.

04

Technology and Regulation

Reviving Purpose Limitation and Data Minimisation in Data-Driven Systems

Asia J Biega* & Michèle Finck**

GDPR, data protection, artificial intelligence, data minimisation, purpose limitation

asia.biega@acm.org

michele.finck@uni-tuebingen.de

This paper determines whether the two core data protection principles of data minimisation and purpose limitation can be meaningfully implemented in data-driven systems. While contemporary data processing practices appear to stand at odds with these principles, we demonstrate that systems could technically use much less data than they currently do. This observation is a starting point for our detailed techno-legal analysis uncovering obstacles that stand in the way of meaningful implementation and compliance as well as exemplifying unexpected trade-offs which emerge where data protection law is applied in practice. Our analysis seeks to inform debates about the impact of data protection on the development of artificial intelligence in the European Union, offering practical action points for data controllers, regulators, and researchers.

1. Introduction

Questions around data management and analysis have been at the fore of policy debates in the European Union in recent years. A particular tension exists between the continued desire to protect personal data through a robust legal regime in the form of the General Data Protection Regulation ('GDPR'), which renders some forms of data collection and analysis unlawful, as well as the objective to generate and analyse more (personal) data so that Europe can remain competitive in the global 'data battle'.[1] There are thus simultaneous policy incentives to process both less and more personal data. This tension will accelerate in the near future with recent legislative developments including the proposed Data Governance Act and the expected Data Act.

This tension can also be pinpointed in relation to debates regarding the legal principles of purpose limitation and data minimisation. While the GDPR has affirmed both principles as core tenets of European data protection law, voices from the private sector, policy circles and academia have argued that these objectives cannot be fulfilled while reaping the benefits of "big data". Our paper examines, through an interdisciplinary law and computer science lens, whether data minimisation and purpose limitation can be meaningfully implemented in data-driven settings, in particular algorithmic profiling, personalisation and decision-making systems. Our analysis reveals that the two legal principles continue to play an important role in managing the risks of personal data processing and that they may even increase the robustness of AI systems by reducing noise in the data. These findings allow us to rebut claims that they have become obsolete.

The paper further highlights that even though these principles are important safeguards in personalisation, profiling, and decision-making systems, there are important limits to their practical implementation. Contrary to what is often claimed, these limits do not so much relate to the quantities of the processed data. Rather, we highlight that the practical difficulties of implementing data minimisation and purpose limitation are due to (A) the difficulties of measuring law and the resulting open computational research questions as well as a lack of concrete guidelines for practitioners; (B) the unacknowledged trade-offs between various GDPR principles, in particular between data minimisation and fairness; (C) the lack of practical means of removing personal data from trained models without considerable economic and environmental costs, and (D) the insufficient enforcement of data protection law.

2. Sources of Disagreement about Purpose Limitation and Data Minimisation

Purpose limitation and data minimisation have been proclaimed to stand in tension with data-driven business models such as those underlying profiling, personalisation and decision-making systems. Arguments against the principles range from technical infeasibility all the way to potentially causing systemic harms to the European economy. At the same time, the principles have been reaffirmed by the GDPR as they limit the collection of unnecessary data in anticipation of potential harms, and aim at maintaining a power balance between

1 Janosch Delcker, 'Thierry Breton: European companies must be ones profiting from European data' (*Politico*, 19 January 2020) https://www.politico.eu/article/thierry-breton-european-companies-must-be-ones-profiting-from-european-data accessed 31 January 2020. Note that the new Commission now sees Europe's competitive edge in industrial rather than personal data.

* Asia J Biega is a tenure-track faculty member and head of the Responsible Computing group, Max Planck Institute for Security and Privacy

** Michèle Finck is Professor of Law and Artificial Intelligence, University of Tübingen. This research was carried out while she was a Senior Research Fellow at the Max Planck Institute of Innovation and Competition, where she remains an Affiliated Fellow.

Received 10 Dec 2020, Accepted 3 Aug 2021, Published 19 Aug 2021.

the data subjects and data controllers. This section provides an overview of those arguments and makes the case for reviving the discussion about these principles in the context of data-driven systems.

2.1 Recent Policy Debates

Our choice to focus on algorithmic profiling, personalisation, and decision-making systems is motivated by two reasons. First, personalisation and profiling have already become key features of many online services and are likely to become even more prominent as an increasing number of online products are accompanied by a service component, a phenomenon referred to as "servitisation".[2] Second, personalisation, profiling and decision-making systems oftentimes use large quantities of data. As such, they are an especially suitable test case to examine the contemporary relevance of data minimisation and purpose limitation.

Personalisation, profiling, and decision-making systems collect personal data in the form of not only user attributes (such as gender or location), but also behavioral interaction logs (such as search queries, product ratings, browsing history, or clicks). The entirety of this data can be used to personalize search ranking results based on past clicks, to personalize product recommendations based on past product ratings, to target ads based on past visited websites, or to make decisions regarding individuals based on topical interest profiles. Thus, a variety of machine learning and data mining setups, including search, recommendation, and classification, fall within the scope of this paper.

Many current uses of machine learning ("ML") in industrial contexts are based on the repurposing of data and legal limitations thereto have been criticized. Mayer-Schönberger and Padova argued that for big data 'to reach its potential, data needs to be gathered at an unprecedented scale whenever possible, and reused for different purposes over and over again'.[3] Voss and Padova noted that 'there is one necessary condition for enabling innovation to flourish: allowing data to be processed without a pre-determined purpose'.[4] According to Moerel and Prins, 'due to social trends and technological developments (such as Big Data and the Internet of Things) the principle of purpose limitation will have to be abandoned'.[5] There are indeed scenarios where the repurposing of data has benefits, such as where speech recordings of voice-operated devices are used to train algorithms seeking to predict information about the health of the speaker.[6]

It has similarly been argued that data minimisation is no longer implementable in settings that generate value from the processing of large quantities of personal data. The incentive in the contemporary data economy is to maximise the accumulation and analysis of personal data. Some consider that with the ubiquitous generation of data the focus should lie on the use rather than the collection of data.[7] Others have pinpointed the dissonance between practices focused on the continuing accumulation of data and the legal principle. Koops has asked: '[w]ho in his right mind can look at the world out there and claim that a principle of data minimisation exists?'[8] Industry organizations have warned that Europe is 'shooting itself in the foot' with limitations on data usage in relation to AI.[9] Others consider that adhering to data minimisation 'would sacrifice considerable social benefit' as it may limit the innovative potential of ML.[10] Yet, data-driven systems present benefits as well as harms and legal intervention can address the latter.

2.2 Computational Evidence

Arguments of non-implementability of data minimisation in contemporary data-driven systems urge an investigation into what computational evidence has to say. On the one hand, the availability of big data has observably enabled progress in machine learning.[11] On the other, we also find evidence demonstrating feasibility of data limitation as well as algorithmic techniques that, in effect, reduce the quantity or the quality of the underlying data.

2.2.1 Minimising the Quantity of Data

Empirical evidence suggests that, in many data-driven settings, using increasingly larger amounts of data leads to diminishing returns in model performance. For example, in 2008, Krause and Horvitz showed that collection of additional user features leads to diminishing returns in the quality of personalized search.[12] Similar trends have since been demonstrated across a variety of ML domains, for instance, in deep learning and its applications ranging from machine translation, through language modeling, to image and speech recognition,[13] in computer vision algorithms,[14] as well as personalised recommendations.[15]

Beyond data retention heuristics focusing on performance-related properties of data, a more straightforward strategy is to retain the most recent data while discarding old data. Research-wise, the effi-

2 Think, for instance, of a "smart" electronic toothbrush connected to an app that offers personalised dental hygiene and toothpaste suggestions to the user.

3 Viktor Mayer-Schönberger and Yann Padova 'Regime Change? Enabling Big Data Through Europe's New Data Protection Regulation' (2016) 17 *The Columbia Science and Technology Law Review* 315, 317.

4 Axel Voss and Yann Padova, 'We need to make big data into an opportunity for Europe' (*Euractiv*, 25 June 2015) https://www.euractiv.com/section/digital/opinion/we-need-to-make-big-data-into-an-opportunity-for-europe accessed 17 January 2020.

5 Lokke Moerel and Corien Prins, 'Privacy for the Homo Digitalis: Proposal for a New Regulatory Framework for Data Protection in the Light of Big Data and the Internet of Things' (25 May 2016) 2 https://papers.ssrn.com/sol3/papers.cfm?abstract_id=2784123 accessed 17 January 2020.

6 International Working Group on Data Protection in Telecommunications, Working Paper on Privacy and Artificial Intelligence, 64th Meeting, 29-30 November 2018, Queenstown (New Zealand), 675.57.14, p 9.

7 Joris van Hoboken, 'From Collection to Use in Privacy Regulation? A Forward Looking Comparison of European and U.S. Frameworks for Personal Data Processing' in Bart van der Sloot, Dennis Broeders, Erik Schrijvers (eds), *Exploring the Boundaries of Big Data* (Amsterdam University Press 2016).

8 Bert-Jaap Koops, 'The Trouble with Data Protection Law' (2014) 4 *International Data Privacy Law* 250, 256.

9 'Artificial Intelligence: How Europe Is Shooting Itself in the Foot with the GDPR' (Fedma) https://www.fedma.org/2018/07/artificial-intelligence-how-europe-is-shooting-itself-in-the-foot-with-the-gdpr accessed 3 December 2020.

10 Mark MacCarthy, 'In Defense of Big Data Analytics' in Selinger et al (eds), *The Cambridge Handbook of Consumer Privacy* (CUP 2018) 56.

11 Alon Halevy, Peter Norvig and Fernando Pereira, 'The Unreasonable Effectiveness of Data' (2009) 24 *IEEE Intelligent Systems* 8.

12 Andreas Krause and Eric Horvitz, 'A Utility-Theoretic Approach to Privacy in Online Services' (2010) 39 *Journal of Artificial Intelligence Research* 633.

13 oel Hestness and others, 'Deep Learning Scaling Is Predictable, Empirically' [2017] arXiv:1712.00409 http://arxiv.org/abs/1712.00409 accessed 9 July 2021.

14 Chen Sun and others, 'Revisiting Unreasonable Effectiveness of Data in Deep Learning Era' [2017] *Proceedings of the IEEE International Conference on Computer Vision* (IEEE 2017) https://openaccess.thecvf.com/content_iccv_2017/html/Sun_Revisiting_Unreasonable_Effectiveness_ICCV_2017_paper.html accessed 25 July 2021.

15 Divya Shanmugam and others, 'Learning to Limit Data Collection via Scaling Laws: Data Minimization Compliance in Practice' [2021] arXiv:2107.08096 [cs] http://arxiv.org/abs/2107.08096 accessed 25 July 2021

cacy of this strategy has been demonstrated in recommender system simulations.[16] One might expect this strategy to perform well especially in settings where user behavior characteristics and preferences change over time and discarding old data might help systems keep user models up to date. One of the recent changes in Google's data retention policy, whereby web activity of new users will by default be deleted after 3 or 18 months,[17] suggests this strategy might be effective in industrial practice as well.

Despite the promise of computational feasibility, data minimisation can lead to unanticipated consequences for both the users and the service providers. Even if limiting quantities of data might lead to little accuracy loss at an aggregate level, studies have shown that data limitation would impact individual users or demographic groups to a different extent, raising the question of what data minimisation might mean in terms of fairness.[18] Minimisation of sensitive attributes has furthermore been shown to hinder the capacity of service providers to audit fairness in personalized products. Last but not least, algorithms exhibit different levels of robustness to data minimisation,[19] raising the question of how limitation obligations would impact different service providers and whether the scale of this impact would depend on how complex or state-of-the-art their algorithms are.[20]

2.2.2 Minimising the Quality of Data

Effects of limiting quantities of data are only one of the sources of disagreement about the desirability of data minimisation. Several studies have shown it is similarly possible to reduce the *quality* of data without reducing its overall quantity. Biega et al.'s simulations demonstrated that it might be possible to achieve good levels of personalisation for search and recommendation while randomly shuffling data (search queries or product ratings) in user profiles under certain accuracy constraints.[21] Similar techniques have been adopted to show the feasibility of such data shuffling techniques in online social communities.[22] Effectively, approaches like these allow a system to retain the volume of data and preserve system accuracy while minimising the quality of aggregated user data profiles.
Other architectures have been proposed in which a user's data resides on their local device while only more crude aggregate data is shared with service providers on a need-to-know basis. The feasibility of such algorithmic architectures has been demonstrated for personalized search (a user shares only high-level topical categories describing their interests with a service provider who personalizes search results)[23] and recommendation (a local tool advises the user whether to share product click and ratings with a recommendation provider based on a privacy-utility analysis).[24] More generally, a local-device distributed learning paradigm called federated learning is an active area of research.[25]

2.2.3 Data-Minimising Algorithmic Techniques

A number of algorithmic techniques *de facto* minimise data. Such techniques include, for instance, outlier detection (for identifying and removing noise and rare anomalies in data), feature selection (for removing features which do not contribute or hurt the learning task), or active learning (for incrementally selecting data to be labelled or added to a model). These strategies were not developed for compliance with the legal principle of data minimisation but rather to help increase the quality of ML models or reduce data acquisition costs. Yet, they do in effect reduce the quantity of data a model uses, demonstrating that, in certain cases, data limitation might result in improved models.

Several recent papers begin to investigate how to adapt these algorithmic techniques to comply with the requirement of data minimisation. Shanmugam et al. propose a framework for automatically learning data collection stopping criteria based on an algorithm's predicted performance curve.[26] The framework adapts to different underlying feature acquisition techniques, including random as well as active learning error-reducing strategies. Goldsteen et al. leverage data anonymisation techniques to suppress and generalise input features in classification.[27] As a result, at the inference stage, a classifier has access to data of reduced quality (feature generalisation), as well as less data overall (feature suppression).

2.3 Benefits and Harms of Data-Driven Systems

The success and acceptance of algorithmic profiling, personalisation, and decision-making systems by both individual users and organizations that develop and deploy them speak to their benefits. Individuals may enjoy an increased quality of digital services, with personalized product recommendations, relevant ads, or search results that surface content satisfying user information needs and effectively helping sift through information overload. More effective profiling may help optimize online marketplaces and help platforms better match content consumers and producers. Profiling may also help organizations with better classification and decision-making. In certain scenarios, where classification and profiling are used to distribute a limited resource, systems may be able to allocate the resource more optimally. On a population level, behavioral data collected through search and online systems could aid developments of societally beneficial solutions for healthcare and well-being improvement, such disease outbreak predictions or detection of disease symptoms.

Personalisation, profiling and decision-making systems are subject to regulatory constraints as they can also result in a range of individual

16 Hongyi Wen and others, 'Exploring Recommendations under User-Controlled Data Filtering' [2018] *Proceedings of the 12th ACM Conference on Recommender Systems* (Association for Computing Machinery 2018) https://doi.org/10.1145/3240323.3240399 accessed 25 July 2021; Asia J Biega and others, 'Operationalizing the Legal Principle of Data Minimization for Personalization' [2020] *Proceedings of the 43rd International ACM SIGIR Conference on Research and Development in Information Retrieval* (Association for Computing Machinery 2020).
17 Google, 'Keeping Your Private Information Private' (*Google Blog*, 24 June 2020) https://blog.google/technology/safety-security/keeping-private-information-private accessed 3 December 2020.
18 Hongyi Wen and others (n 16); Asia J Biega and others (n 16.
19 Asia J Biega and others (n16).
20 Xavier Amatriain, 'In Machine Learning, What Is Better: More Data or Better Algorithms' https://www.kdnuggets.com/2015/06/machine-learning-more-data-better-algorithms.html accessed 8 July 2021.
21 Asia J Biega, Rishiraj Saha Roy and Gerhard Weikum, 'Privacy through Solidarity: A User-Utility-Preserving Framework to Counter Profiling', [2017] *Proceedings of the 40th International ACM SIGIR Conference on Research and Development in Information Retrieval* 675.
22 Sedigheh Eslami and others, 'Privacy of Hidden Profiles: Utility-Preserving Profile Removal in Online Forums' [2017] *Proceedings of the 2017 ACM on Conference on Information and Knowledge Management* 2063
23 Xuehua Shen, Bin Tan and ChengXiang Zhai, 'Privacy Protection in Personalized Search' (2007) 41 *ACM SIGIR Forum* 4.
24 Rachid Guerraoui, Anne-Marie Kermarrec and Mahsa Taziki, 'The Utility and Privacy Effects of a Click' [2017] *Proceedings of the 40th International ACM SIGIR Conference on Research and Development in Information Retrieval* 665.
25 Tian Li and others, 'Federated Learning: Challenges, Methods, and Future Directions' (2020) 37 *IEEE Signal Processing Magazine* 50..
26 Divya Shanmugam and others (n 15).
27 Abigail Goldsteen and others, 'Data Minimization for GDPR Compliance in Machine Learning Models' [2020] arXiv:2008.04113 [cs] http://arxiv.org/abs/2008.04113 accessed 27 July 2021.

and societal harms.[28] The storage of data in those systems poses privacy risks because of potential security breaches or inadequate technical and organisational measures adopted by the data controller. For instance, if an anonymized version of the data is shared,[29] identity linking is still possible because perfect anonymisation is often infeasible.[30] Moreover, the richness of observations about a whole population of users often enables inference of additional information about individuals that is not explicitly present in the data. For this reason, it is challenging to assess the implications of data processing *ab initio*. Feasibility of inference for attributes including political convictions, sexual preferences, or personality traits has been demonstrated for social media data[31], movie rating data[32], query suggestions or targeted ads.[33] Systems might moreover make incorrect inferences about a person because of the inherent system inaccuracy or when more than one person uses the same device or account.[34] Incorrect inferences may lead to a range of consequences, from user embarrassment, to unfair denial of opportunities. Indeed, due to the various risks of personalised behavioral advertising, for instance, some have called for bans.[35]

Further harms arise because of a high complexity and a lack of transparency of data-driven systems. Many users do not understand how the systems work or what happens to their data[36], which might result in a loss of control or a sense of helplessness and powerlessness. Further feelings of unease might stem from perceptions of surveillance and a loss of privacy.[37]

On a societal level, risks of profiling and personalisation include increased surveillance, targeted censorship in authoritarian regimes, or filter bubbles.[38] Profiling can moreover reinforce 'different forms of social, cultural, religious, legal and economic segregation and discrimination' and enable the microtargeting of individuals in a manner that may profoundly affect their lives. The fact that optimisation processes inevitably prioritise certain values over others shapes online environments in a manner that can be detrimental. Beyond, ML is considered to 'influence emotions and thoughts and alter an anticipated course of action, sometimes subliminally' which may affect not only economic choices but also social and political behaviours, particularly if used without democratic oversight or control. Sub-conscious and personalised levels of algorithmic persuasion may moreover have significant effects on the cognitive autonomy of individuals and their right to form opinions and take independent decisions.[39] Recital 75 GDPR explicitly recognises that personal data processing carries risks, particularly where it serves to create 'personal profiles'.

There are moreover major risks inherent in the business model that underlies personalised advertising. Hwang has pointed out that most 'free' online services are currently financed by advertising revenue.[40] However, it is possible that personalised advertising may have little to no benefit compared to non-personalised alternatives.[41] The realisation that personalised advertising is not, in fact, superior to non-personalised alternatives such as contextualized advertising or other events such as an economic crisis may lead to a withdrawal of income for online service providers, which as a consequence no longer have a means of sustaining their operations. This would dramatically affect online services as we know them. These are more than hypothetical risks in the aftermath of a global pandemic that may trigger economic depression and result in a dramatic cut in corporate advertising budgets, particularly in light of increasing evidence that the benefits of personalised advertising hardly outweigh non-personalised alternatives.[42] Seen from this perspective, reliance on personalisation equals systemic risk.

3. Purpose Limitation

In essence, purpose limitation requires that the data controller define *ab initio* the purpose(s) for which personal data will be processed. This pre-defined purpose should not be exceeded, save where the new purpose is sufficiently approximate to the initial purpose or where there is an additional legal basis for further processing such as data subject consent or the need to process data for purposes such as scientific research. Purpose limitation is as old as data protection law itself and essentially serves the goal of minimising the risks that arise where personal data is processed in confining the possibilities of its usage by limiting instances of lawful processing. This section introduces the purpose of personal data processing from a legal, practical, and computer science perspective.

3.1 The Legal Obligation to Define a Purpose

Article 8(2) of the Charter of Fundamental Rights provides that 'data must be processed fairly for specified purposes'[43] and according to Article 5(1)(b) GDPR personal data shall be:

28 See also Orla Lynskey, 'Track[ing] changes: an examination of EU Regulation of online behavioural advertising through a data protection lens' (2011) 36 *European Law Review*, 874, 879-881.
29 Michael Barbaro and Tom Zeller Jr, 'A Face Is Exposed for AOL Searcher No. 4417749' (*The New York Times* 9 August 2006) https://www.nytimes.com/2006/08/09/technology/09aol.html accessed 3 December 2020.
30 Michèle Finck and Frank Pallas, 'They Who Must Not Be Identified - Distinguishing Personal from Non-Personal Data Under the GDPR' (2020) 10 *International Data Privacy Law*, 11-36.
31 Sibel Adali and Jennifer Golbeck, 'Predicting Personality with Social Behavior' [2012] *2012 IEEE/ACM International Conference on Advances in Social Networks Analysis and Mining* 302; Michal Kosinski, David Stillwell and Thore Graepel, 'Private Traits and Attributes Are Predictable from Digital Records of Human Behavior' (2013) 110 *Proceedings of the National Academy of Sciences* 5802.
32 Arvind Narayanan and Vitaly Shmatikov, 'Robust De-Anonymization of Large Sparse Datasets' [2008] *2008 IEEE Symposium on Security and Privacy* (sp 2008) 111.
33 European Network and Information Security Agency (ENISA), 'Privacy considerations of online behavioural tracking' (2012) 13-14 https://www.enisa.europa.eu/publications/privacy-considerations-of-online-behavioural-tracking/at_download/fullReport accessed 31 January 2020.
34 Tara Matthews and others, '"She'll Just Grab Any Device That's Closer": A Study of Everyday Device & Account Sharing in Households' [2016] *Proceedings of the 2016 CHI Conference on Human Factors in Computing Systems* 5921.
35 Edelamn Gilad, 'Why Don't We Just Ban Targeted Advertising?' (*WIRED*, 22 March 2020) https://www.wired.com/story/why-dont-we-just-ban-targeted-advertising accessed 3 December 2020.
36 Catherine Miller, Rachel Coldicutt and Hannah Kitcher, 'People, Power and Technology: The 2018 Digital Understanding Report' (Doteveryone 2018) http://understanding.doteveryone.org.uk accessed 3 December 2020.
37 Orla Lynskey (n 28) 874, 879-881.
38 European Network and Information Security Agency (ENISA) (n 33).
39 Council of Europe, Declaration by the Council of Ministers on the Manipulative Capabilities of Algorithmic Processes (13 February 2019) https://search.coe.int/cm/pages/result_details.aspx?ObjectId=090000168092dd4b
40 Tim Hwang, Subprime Attention Crisis (FSG Originals X Logic 2020).
41 Natasha Lomas, 'Targeted Ads Offer Little Extra Value for Online Publishers, Study Suggests' (*TechCrunch*, 31 May 2019) https://techcrunch.com/2019/05/31/targeted-ads-offer-little-extra-value-for-online-publishers-study-suggests accessed 3 December 2020.
42 Ster Reclame, 'Online advertising 2.5 years after the implementation of the GDPR: what are the lessons learned? (*Ster.nl*, 8 December 2020) https://www.ster.nl/online-advertising-2-5-years-after-the-implementation-of-the-gdpr-what-are-the-lessons-learned accessed 12 August 2021.
43 Charter of Fundamental Rights of the European Union [2000] OJ C 364/1, art 8(2).

collected for specified, explicit and legitimate purposes and not further processed in a manner that is incompatible with those purposes; further processing for archiving purposes in the public interest, scientific or historical research purposes or statistical purposes shall, in accordance with Article 89(1), not be considered to be incompatible with the initial purposes

Article 5(1)(b) lists two distinct components: (i) purpose specification and (ii) compatible use. Purpose specification requires that personal data should only be collected for 'specified, explicit and legitimate purposes' whereas compatible use mandates that personal data shall not be 'further processed in a manner that is incompatible with those purposes.'[44] Ultimately, purpose limitation serves to manage the risk that inevitably arises when personal data is processed.

3.1.1 Purpose Specification

Pursuant to Article 5(1)(b) GDPR, the data controller must (i) communicate the purposes for which data is processed, which must be (ii) explicit and (iii) legitimate. This forces controllers to precisely define what data they need and discourages the accumulation of personal data for speculative future use. Importantly, the specification ought to occur before data collection (or any other processing) starts.[45]

Purpose specification can be broken down into three distinct requirements. First, the purpose must be *specified*, meaning that it must be sufficiently precise to enable the implementation of data protection safeguards and be useful to the data subject.[46] The Article 29 Working Party considers that general statements such as 'improving user experience', 'for commercial purposes' or 'for advertising' are generally not specific enough.[47] This finding is important as personalisation, profiling, and decision-making algorithms generally process personal data seeking to improve rather than simply provide a service, a description that fails the specificity test. Yet, if, as per the Working Party, improvements of user experience are not a valid purpose, it is worth wondering whether improvements in service delivery can ever be.

Such definitions are, however, contextual as the level of detail required will depend on the specific context.[48] In scientific research broader formulations are permissible as it is often not possible to fully identify the purpose at the time of data collection.[49] National supervisory authorities have in the past taken enforcement action against definitions of purpose considered to be insufficiently specific. In 2014, the Dutch DPA imposed a cease and desist order on Google, arguing that 'the provision of the Google service' was not specific enough.[50] In 2020, the Spanish supervisory authority fined a bank for violation of purpose limitation when it processed customer data for sixteen years after the end of the corresponding business relationship.[51] In 2021, the Belgian supervisory authority held that a school's parental mailing list, which did not make use of the blind carbon copy ('BCC') function, breached Article 5(1)(b) GDPR as it was not necessary to circulate all parent emails to achieve the informational purposes.[52] The literature has also drawn attention to violations of purpose specification in online advertising.[53]

Data subjects' expectations must also be accounted for. In principle, this is a laudable perspective as it takes into account the interests of the data subject. Yet, the principle's usefulness can also be questioned as data subjects' understanding of contemporary data ecosystems is extremely limited.[54] One may also wonder whether over time, the principle eradicates its own usefulness. As data collection and use practices change, so do expectations. Many current practices would likely not have been acceptable in the 1990s, whereas in the future people might be accepting of practices that would now be seen as crossing a red line. More extreme data processing may thus ultimately result in more acceptance thereof.

Second, the purpose must be *explicit*, meaning that it 'must be sufficiently unambiguous and clearly expressed'. This requires that the purposes 'must be clearly revealed, explained or expressed in some intelligible form'.[55] Where this is not the case, factual elements, common understandings and reasonable expectations are considered to determine the actual purpose.[56] Whereas purposes must be specifically defined, they should also be understandable to data subjects. To achieve both ends, layered notices are encouraged as they can both provide an overall explanation and sufficient granularity.[57] Requiring information to be explicit also underlines the connection between the purpose limitation and transparency principle, according to which data subjects must be provided with 'concise, transparent, intelligible and easily accessible' information about personal data processing.[58]

Third, the purpose ought to be *legitimate*. Legitimacy mandates that processing occurs in line with applicable law such as non-discrimination, criminal or employment law.[59] All elements of EU and national law (including municipal decrees and case law) must be respected. Legitimacy may also require respect of 'customs, codes of conduct, codes of ethics[60], contractual arrangements and the general context and facts of the case'.[61] Whereas reliance on such elements would depend on context, this is an interesting point in particular in light of the spread of AI 'ethics codes', which are designed as non-binding instruments.

The information to be provided is contextual: a small shop does not need to provide as much detail as a transnational company.[62] Where a broad user group across different cultures is targeted, information needs to be particularly clear and where a controller provides different services (such as email, social networking and photograph, video and music uploads) granularity is needed to make sure the information provided is sufficiently clear.[63] Where services are offered to particular groups such as the elderly or asylum applicants, their specific charac-

44 Ibid, 3-4.
45 Article 4(2) GDPR adopts a broad definition of 'processing' to include 'any operation or set of operations which is performed on personal data or on sets of personal data'.
46 GDPR, recital 39. See also Article 29 Working Party, Opinion 03/2013 on purpose limitation (WP 203) 00569/13/EN, 15-16.
47 Article 29 Working Party, Opinion 03/2013 on purpose limitation (WP 203) 00569/13/EN, 16.
48 Ibid.
49 GDPR, recital 33.
50 Joris van Hoboken (n 7).
51 https://gdprhub.eu/index.php?title=AEPD_-_PS/00076/2020.
52 https://gdprhub.eu/index.php?title=APD/GBA_-_03/2021.
53 https://hal.inria.fr/hal-02566891/document
54 Catherine Miller, Rachel Coldicutt and Hannah Kitcher (n 36).
55 Article 29 Working Party, Opinion 03/2013 on purpose limitation (WP 203) 00569/13/EN, 17.
56 Ibid, 19.
57 Ibid, 16.
58 GDPR, art 12(1) and recital 58 .
59 Article 29 Working Party, Opinion 03/2013 on purpose limitation (WP 203) 00569/13/EN, 12.
60 For a critique of favoring ethics codes over law, see http://ejlt.org/article/view/722/978#_ednref26
61 Article 29 Working Party, Opinion 03/2013 on purpose limitation (WP 203) 00569/13/EN, 20.
62 Ibid, 51.
63 Ibid, 51.

teristics need to be accounted for.[64]

It is finally worth noting that personal data can be collected for more than one purpose. Where these purposes are related, an 'overall purpose' can be used (under whose umbrella a number of separate processing operations take place), yet controllers must be careful not to identify a broad purpose in view of undertaking further processing that is only remotely related to the initial purpose.[65]

Purpose limitation is an essentially procedural requirement which does—with the exception of the legitimacy requirement—not appear to have a substantive facet. Seen from this perspective, it would mainly be an exercise in skilled legal drafting as any legitimate purpose that is formulated with sufficient specificity and explicitness would be GDPR-compliant. It is for instance worth wondering whether in the Belgian school email case referenced above, the school could have sent such an email had it defined the purpose not just as information but also networking between parents. Seen from this perspective, purpose limitation is a mindfulness exercise for data controllers in that it obliges them to ponder the use of personal data and be explicit about what it is being used for. It creates a reasonableness requirement for personal data usage in that the definitions of purpose that are too broad, including (as seen above) 'improving user experience', which is what many algorithmic personalisation, profiling and decision-making systems do, is too broad. However, if, say, a video streaming provider were to use better-skilled legal drafting to state that the purpose of processing is to 'provide personalized video recommendations', that would very likely be satisfactory. It follows that if skilled legal drafting can define the purpose in reasonable ways, it will likely pass the purpose specification test. As such, purpose specification does not genuinely limit the ways in which personal data can be used.

Both the computational and practical perspectives have revealed that personalisation and profiling usually serve to improve the service that is delivered. As per existing regulatory guidance, this is not a specific enough purpose. In order for such practices to become aligned with data protection law, they need to be more *specific* which can be achieved by more detailed, or layered statements. Yet, if more precise language is all that is needed to ensure compliance with purpose specification, it is worth wondering what objective it actually serves in data protection law. Can any purpose be used (as long as legitimate and explicit) provided that it is put in precise language? If so, what objective does purpose specification actually fulfil? This would indicate that any purpose can be realised as long as it is formulated specifically and corresponds to other data protection requirements.

3.1.2 Compatible Use

The second component of purpose limitation is compatible use. It requires that personal data be not further processed in a manner incompatible with the original purpose(s). However, the mere fact that data is processed for a purpose different from that originally defined does not mean that it is automatically incompatible.[66] In some circumstances, processing for a different purpose is considered sufficiently connected to the original purpose. This requires a case-by-case evaluation of whether the initial and further processing are compatible. Here, relevant criteria according to the Article 29 Working Party's 2013 guidance are (i) the relationship between the different purposes; (ii) the context of collection and the reasonable expectations of data subjects; (iii) the nature of personal data and the impact of further processing on data subjects; and (iv) the safeguards adopted by the controller.[67] National DPAs have also issued guidance on the interpretation of compatible use. The UK Information Commissioner's Office ('ICO') for instance considers that the new use must be 'fair, lawful and transparent'.[68]

To assess whether further processing was implied in the original purpose, adopting the perspective of 'a reasonable person in the data subject's position' has been recommended.[69] However, as observed above, consumers rarely have a realistic understanding of contemporary processing practices. Moreover, the nature of the contract and the relation between the data subject and the data controller are to be accounted for.[70] For example, the public disclosure of personal data is a relevant factor.[71]

Processing that is incompatible with the original purpose cannot be legitimized through reliance on an alternative legal ground under Article 6 GDPR.[72] In principle, data controllers would thus have to anonymise personal data to process it beyond the limited purpose (as this would bring the data outside of the scope of the GDPR).[73] This is, however, often easier said than done considering the difficulties of achieving anonymisation.[74] The compatible use requirement is a much stronger practical constraint on data processing than purpose specification, which, as seen above, is largely an exercise in skilled legal drafting. Once the purpose has been specified, however, compatible use does impose considerable practical constraints on the possibilities of data use. This is a general challenge in respect of the European Commission's current policy agenda that seeks to further incentivise the sharing of data.[75] In our context, it for instance applies that where the purpose of the collection of an address is specified as necessary for billing purposes, this information cannot be used to inform personal recommendations on the basis of location. There are, however, a number of additional instances where data can be processed beyond the purpose.

A question remains of when two purposes are compatible. When the processing purpose is stated to be an improvement in performance and the performance is measured using well-defined metrics, one natural computational interpretation of compatible use is when metrics are positively correlated. Consider the following example. An online outdoors store originally collected personal data such as product ratings to improve the performance of personalized hiking gear recommendations. The store expands its catalogue to include hiking clothing, and it turns out that shopper preferences for certain categories of clothing and certain categories of gear are correlated (e.g., producer brand, price range, or other features used for person-

64 Ibid.
65 Ibid, 16.
66 Ibid, 21.

67 Ibid, 23-26.
68 Information Commissioner's Office, 'Guide to the General Data Protection Regulation (GDPR) – Principle (b): Purpose limitation' https://ico.org.uk/for-organisations/guide-to-data-protection/guide-to-the-general-data-protection-regulation-gdpr/principles/purpose-limitation accessed 10 January 2020.
69 Article 29 Working Party, Opinion 03/2013 on purpose limitation (WP 203) 00569/13/EN, 24.
70 Ibid.
71 Ibid, 26.
72 Article 29 Working Party, Opinion 03/2013 on purpose limitation (WP 203) 00569/13/EN, 3.
73 Ibid, 7.
74 Luc Rocher, Julien Hendrickx and Yves-Alexandre de Montjoye, 'Estimating the success of re-identifications in incomplete datasets using generative models' (2019) 10 *Nature Communications* 3069 and Michèle Finck and Frank Pallas (n 30).
75 Such as through the proposed Data Governance Act.

alisation). The data lawfully collected for the purpose of improving personalized gear recommendation will likely improve the clothing recommendations, thus it could be argued that it can be used for the latter purpose under compatible use. Practical implementations would need to account for various challenges, including the existence of spurious correlations or handling of implicit latent features in the algorithms.

3.2 Current State

To understand the purpose limitation *status quo*, we consider how key service providers using ML define the purpose of personal data processing. Google has a lengthy layered description. It first informs users that data is collected 'to build better services' before providing examples of what this means.[76] In relation to personalisation, information is used to provide services such as 'recommendations, personalized content, and customized search results' as well as personalized ads (depending on a user's settings). Google further informs its users that it uses 'automated systems that analyze your content' to provide customized search results or ads. Facebook also adheres to a layered approach by informing users that personal data is used to 'provide and support the Facebook Products', which includes 'to personalise features and content (...) and make suggestions for you (...) on and off our Products'.[77] Netflix informs its users that it processes personal data to: (i) receive newsletters, (ii) send push notifications, (iii) enhance customer experience, and (iv) fulfil legal or contractual obligations.[78] These are but some examples that highlight that the purpose of data processing can be defined in a number of ways, in line with the service provider's product and objective. As will be seen below, a weakness of the current system is that, despite publically available data processing policies and development of automated tools that can analyze them,[79] there are virtually no means of checking whether these verbal expressions correspond to what happens in practice, highlighting the difficulties of practically enforcing data protection law.

3.3 Service Improvement as Purpose Specification

Service providers state that user interaction data is collected to provide, improve, or personalize their services. Yet, some of these purposes are in fact not well-grounded in computational practice, as it is not immediately clear whether and which data is actually necessary to improve service results.

Firstly, it is crucial to observe that, from a system's perspective, ongoing collection of user data is not necessary to *provide* services such as search, recommendation, or classification. In web search, a ranking of webpages can be computed by matching keywords in queries to words appearing in web pages. In fact, ranking methods based on properly weighted word statistics beat some of the more complex methods in search benchmarks.[80] Moreover, while processing a query might be necessary to complete a given search transaction, it might not be necessary to store it for future use once the search task is complete. In their search personalisation audit studies, Hannak et al. have not observed search-history-based personalisation in Google or Bing.[81] Recommendations can be unpersonalized and based on external data (for instance, sales statistics of cinema tickets could be used as a popularity indicator for recommending movies) or even random. Thus, *providing* a service in such scenarios does not appear to be a valid purpose for collection of user data. Instead, user data in personalisation, profiling and decision-making systems is collected to *improve* the results. Service improvement could be considered an objective criterion for purpose formulation if it were legitimate, explicit and specific enough.

Determining whether improvement of a service is legitimate could be thought of as conditional on the legitimacy of the service itself. In most cases, improving a legitimate service might be considered legitimate as well.

Explicitness, as argued by Koops,[82] could be enhanced by specifying purposes in a machine-readable format such as XML. Indeed, such code-driven expression forces data controllers to reflect on their processing goals explicitly. Fouad et al. has proposed to improve explicitness of browsing cookie purposes by listing them in a structured table in the data processing policy.[83] This approach thus allows for quick identification of processing purposes. A solution along these lines appears viable for data collection purposes in data-driven systems as well.

As for specificity, however, our earlier analysis revealed the question of whether service improvement can be considered a purpose specific enough (since, as per the Working Party, improvement of user experience is not).[84] We thus consider ways in which improvement can be stated more concretely to pass the specificity test.

Von Grafenstein has argued that purpose standardisation aids in increasing purpose specificity and legal certainty, as 'both the individuals concerned and data controllers, which are part of this "purpose"-oriented system, are reassured that all data processing occurs under the same conditions.'[85] To this end, Fouad et al. proposed using ontologies not only to standardize purpose descriptions but also to allow reasoning about ontological relations between purposes, such as subsumption.[86]

In the context of data-driven systems, relations between improvement purposes could be defined along the axes of *what* and *how*. Increasing specificity along the *what* axis might entail defining which functionality in the system would exactly be improved. For instance, a layered description might indicate improvements in personalized search, and within that purpose, specify which topics of search queries will see improvement in the results. Increasing specificity along the *how*

76 Google, 'Privacy Policy' (15 October 2019) https://policies.google.com/privacy?fg=1#whycollect accessed 17 January 2020.

77 Facebook, 'Data Policy' (19 April 2019) https://en-gb.facebook.com/privacy/explanation accessed 17 January 2020.

78 https://www.whats-on-netflix.com/privacy-policy Accessed 3 December 2020.

79 Shomir Wilson and others, 'Analyzing Privacy Policies at Scale: From Crowdsourcing to Automated Annotations' (2019) 13 *ACM Transactions on the Web* 1; Abhilasha Ravichander and others, 'Question Answering for Privacy Policies: Combining Computational and Legal Perspectives' [2019] *Proceedings of the 2019 Conference on Empirical Methods in Natural Language Processing and the 9th International Joint Conference on Natural Language Processing* (EMNLP-IJCNLP) 4947.

80 Nick Craswell and others, 'Overview of the TREC 2019 Deep Learning Track' [2020] arXiv:2003.07820 [cs] http://arxiv.org/abs/2003.07820 accessed 2 December 2020.

81 Aniko Hannak and others, 'Measuring Personalization of Web Search' [2013] *Proceedings of the 22nd international conference on World Wide Web - WWW '13* 527.

82 Bert-Jaap Koops, 'The (In)Flexibility of Techno-Regulation and the Case of Purpose-Binding' (2011) 5 *Legisprudence* 171, 186

83 Imane Fouad and others, 'On Compliance of Cookie Purposes with the Purpose Specification Principle' [2020] *2020 IEEE European Symposium on Security and Privacy Workshops* (EuroS PW) 326.

84 Article 29 Working Party, Opinion 03/2013 on purpose limitation (WP 203) 00569/13/EN, 16.

85 Max von Grafenstein, The Principle of Purpose Limitation in Data Protection Laws (Nomos Verlagsgesellschaft mbH & Co KG 2018) 645

86 Imane Fouad and others (n 83) 331

axis might entail specifying how improvement will be quantified. For instance, the goal of a system might be to display relevant documents a certain number of ranking positions higher in the search results than they would be without the collected personal data. Neither of the suggested directions are straightforward to implement, however.

3.4 Computational Challenges

Improvement is an ambiguous concept and needs additional specification in practical and computational terms. Generally, it is reasonable to assume improvements would be quantified as *differences* in selected *system performance metrics*. Performance evaluation is widely used to judge models and systems in scientific publications, to measure progress in the field through public benchmarks[87], or to determine if updates to tech products should be shipped using techniques such as A\B testing[88]. A natural consequence would thus be to similarly reason about purpose limitation via performance and *tie the purpose of data collection to improvements in system performance*. Practically, to form a quantitative basis of purpose, it remains to be determined (i) which metrics to choose, (ii) how to obtain their values, and (iii) which level to aggregate metric differences at.

A metric would have to be selected from a suite of metrics that often guide system quality measurement. The main reason for such complex evaluation setups is that different metrics capture different aspects of performance, and often none of these aspects is more important than another. Moreover, individual metrics in a suite will often disagree as to whether a change leads to service improvement. Further complications include the fact that metrics might differ by application domain. A system performance will be measured differently for personalized movie recommendations than for personalized search. Finally, it is important to acknowledge that metrics often serve as simpler, quantifiable proxies for measurement targets. For instance, in systems such as search and recommendation, the goal might be to improve 'user satisfaction'. User satisfaction is, however, approximated using simpler measurable concepts such as the number of clicks.

Despite the fact that metrics are imperfect approximations of hard-to-model concepts such as 'user satisfaction', using them as a ground for purpose limitation would enable proxy metrics to determine which data should and should not be collected. Barocas and Selbst discuss the caveats behind a related concept of target variables in machine learning models deployed in societally sensitive applications.[89] Values of performance metrics can be obtained using quantitative, qualitative or mixed evaluation methods. For instance, in personalized search the goal of a system might be to reduce the time necessary to find the desired information for ambiguous queries. Whether the system achieves this goal might be measured quantitatively using the rate of query rephrasing (when a user rephrases their query, the preceding query has likely not yielded satisfying results), or qualitatively, through an in-person user experience interview.

Finally, there might be multiple approaches to aggregating metric improvements. Among many possible options, a formal definition might require that the collection of data improves the service on average for all users, or that the collection of data improves the service for the individual from whom the data is collected.[90]

More detailed guidance will be necessary for practitioners to navigate all the discussed design choices as they are likely to lead to different minimisation outcomes.

4. Repurposing Personal Data Beyond Compatible Use

The purpose limitation principle requires that before any personal data processing can take place the purpose thereof be defined. Subsequent processing is assessed against that purpose. This stands in contrast with the reality of much contemporary data mining practice where the data that is mined was often initially collected for another purpose. Beyond the compatible use requirement examined above, the GDPR acknowledges other avenues of processing data beyond the initial purpose. First, the scientific research exemption recognises that in this specific context, it is often difficult to foresee potential future uses of personal data.[91] Second, the GDPR acknowledges that where personal data is further processed for 'statistical purposes', this shall not be considered incompatible with the original purpose.[92] Third, and more controversially, data controllers are also able to move beyond purpose limitation in getting data subjects to consent to further processing.

4.1 Scientific Research

Article 5(1)(b) GDPR foresees that personal data can also be further processed for 'scientific' purposes in accordance with the safeguards listed in Article 89 GDPR, including technical and organizational measures and respect for data minimisation. In such circumstances, member state law may also provide for derogations from some data subject rights.[93] The scientific purpose exemption shall be 'interpreted in a broad manner including for example technological development and demonstration, fundamental research, applied research and privately funded research'.[94]

Given the broad definition of scientific research it is pertinent to wonder what can be considered to be 'scientific research' in the context of personalisation, profiling, and decision-making systems. While publicly funded and externally published work at academic institutions might rather uncontroversially be considered as research, what constitutes research at private technology companies is less clear. Industrial research teams might do both internally and externally facing work, with the internal research not meant for external publication but rather for proprietary product development and innovation. To complicate matters further, various company organizational structures often make it impossible to distinguish who works on research and who works on products, despite official employee titles that might suggest a clear distinction. For example, at Google, research scientists are embedded in engineering teams[95] and as a result many research scientists develop products, and many software engineers

87 Benchmarks are at the center of some Computer Science conferences such as TREC https://trec.nist.gov/overview.html (accessed 24 February 2020), or TAC KBP https://tac.nist.gov/about/index.html (accessed 24 February 2020), both organized by the National Institute of Standards and Technology.

88 Ron Kohavi and Roger Longbotham, 'Online Controlled Experiments and A/B Testing' in Claude Sammut and Geoffrey I Webb (eds), *Encyclopedia of Machine Learning and Data Mining* (Springer US 2017) http://link.springer.com/10.1007/978-1-4899-7687-1_891 accessed 3 December 2020.

89 Solon Barocas and Andrew D Selbst, 'Big Data's Disparate Impact' (2016) 104 *California Law Review* 671.

90 Asia J Biega and others (n 16).

91 GDPR, arts 5(1)(b) and 89.

92 See also GDPR, recital 50: 'processing for archiving purposes in the public interest, scientific or historical research purposes or statistical purposes should be considered to be compatible lawful processing operations'.

93 Article 89(2) GDPR.

94 Recital 159 GDPR.

95 Alfred Spector, Peter Norvig and Slav Petrov, 'Google's Hybrid Approach to Research' (2012) 55 *Communications of the ACM* 34.

do externally-facing research. Spotify does not follow traditional organizational hierarchies, and employees with different backgrounds and titles are organized according to various functional dimensions, for instance, system feature areas.[96]

4.2 Statistical Purposes

The GDPR moreover permits further processing for another purpose where that purpose is 'statistics.'[97] If an activity qualifies as statistical analysis, controllers benefit from a more favorable regime, including that data can be kept for longer than necessary for the purposes of processing.[98] The recognition of a more favorable regime for statistical analysis reflects that traditionally, data used for statistics was usually initially collected for another purpose. For example, national statistics offices have relied on data collected for other ends to carry out their work. Given the overlaps between statistics and computational learning, it is worth enquiring whether ML can be qualified as statistical analysis under the GDPR to benefit from the corresponding legal regime. Indeed, just as statistics, data used to train ML systems is often repurposed.

Recital 162 GDPR defines statistical purposes as a form of processing 'necessary for statistical surveys or for the production of statistical results.' These results may be further used for different purposes. The recital, however, also makes clear that the output must not be 'personal data, but aggregate data' and that, moreover, the results 'are not used in support of measures or decisions regarding any particular natural person'.[99] The GDPR is thus clear that some ML outputs cannot be qualified as 'statistics', namely those that generate personal data or are used to support individual measures and decisions.[100] It seems uncontroversial that personalised services are indeed an individual measure.

From the above it would seem that some forms of ML output could qualify as statistics whereas others cannot. Concretely, whereas the prediction of overall customer churn would be statistics, the prediction of whether a given customer will leave and the initiation of corresponding action (such as more attractive personalised deals) could not fall within the scope of this more favorable regime. The mere fact that statistical methods are used in private commercial settings (as opposed to statistical analysis in the public interest) nonetheless does not form a bar to the application of the statistical exemption, which does apply to 'analytical tools of websites or big data applications aimed at market research'.[101] The GDPR furthermore enables the EU or Member States to create specialised regimes on processing personal data for statistics.[102] If one or several Member States would choose this route (such as to attract data analysis companies to their jurisdiction) there is a clear risk of fragmentation in the Digital Single Market - going counter the GDPR's harmonising objective.

Although computing experts disagree whether machine learning is different from statistics, many do point out that they are indeed different to a certain extent. Some argue that they have different goals (prediction vs. inference and analysis of relations between random variables) while sharing some of the algorithms and practices.[103] [104] Others argue that the disciplines are complementary although increasingly converging.[105]

There is another distinction relevant in the context of the above discussion on individual vs. aggregate outputs. Namely, machine learning pipelines produce aggregate and individual results at different processing stages. Many of the systems considered in this article leverage large sets of user data to construct a model, and then use such an aggregate model in conjunction with an individual's data to compute the individual's results. For instance, a search engine might train a non-personalized ranker that preselects webpages as a response to a query, and then use an individual's personal data to re-rank the webpages in the preselected set. In a scenario like this, the training of the aggregate model might be considered a form of statistical analysis (and thus not subject to data minimisation), while applying the model in conjunction with an individual's data will not (as it produces individual results).

One caveat to consider is that there exist personalisation algorithms that do not conform to the above scheme. For instance, in personalized recommendation models based on matrix factorisation techniques, all individuals' data is used to train the model and no new data is used at the application stage. In such a case, a given subject's data is used both to train an aggregate model as well as to produce individual results.

4.3 Consent as the Silver Bullet?

Where processing goes beyond compatible use, it can be legitimized by the data subject's consent (or where it is based on EU or Member State law).[106] Subject to consent, 'the controller should be allowed to further process the personal data irrespective of the compatibility of the purposes'.[107] Today, many data controllers use consent to legitimise data processing for purposes that would otherwise not be lawful as they exceed the initial purpose.[108] Where ML is used to inform measures or decisions in relation to individuals, consent 'would almost always be required', in particular for direct marketing, behavioural or location-based advertisement, data-brokering, or tracking-based digital market research.[109] There accordingly appears to be an assumption that these types of analysis are too different from the original purpose to be legitimised by compatible use.

From this perspective, consent appears as the silver bullet to get around legal limitations of purpose limitation. However, using consent to legitimise otherwise illegitimate data processing has been

96 Atlassian, 'The Spotify Model' (Atlassian) https://www.atlassian.com/agile/agile-at-scale/spotify accessed 3 December 2020.
97 GDPR, art 5(1)(b).
98 GDPR, art 5(1)(e).
99 GDPR, recital 162.
100 It is true that this is provided in the recital and not the text of the GDPR itself. A contrary conclusion of a court would, however, be surprising, considering the ECJ's general approach to recitals.
101 Article 29 Working Party, Opinion 03/2013 on purpose limitation (WP 203) 00569/13/EN, 29.
102 Recital 162 provides that they can 'determine statistical content, control of access, specifications for the processing of personal data for statistical purposes and appropriate measures to safeguard the rights and freedoms of the data subject and for ensuring statistical confidentiality'.
103 Tom Fawcett and Drew Hardin, 'Machine Learning vs. Statistics' (*Silicon Valley Data Science*, 10 August 2017) https://www.svds.com/machine-learning-vs-statistics accessed 12 August 2021.
104 Danilo Bzdok, Naomi Altman and Martin Krzywinski, 'Statistics versus Machine Learning' (2018) 15 *Nature Methods* 233.
105 Max Welling, 'Are ML and Statistics Complementary?' (2015) https://www.ics.uci.edu/~welling/publications/papers/WhyMLneedsStatistics.pdf accessed 3 December 2020.
106 GDPR, art 21 and recital 50. Note, however, that the data controller must safeguard the right to object.
107 GDPR, recital 50 .
108 See, by way of example, Google, 'Privacy Policy' (15 October 2019) https://policies.google.com/privacy?fg=1#whycollect accessed 17 January 2020.
109 Article 29 Working Party, Opinion 03/2013 on purpose limitation (WP 203) 00569/13/EN, 46.

criticised.[110] Consumer rights organizations have pointed out that it allows data controllers to circumvent purpose limitation and makes it hard for consumers to understand contemporary data flows.[111] In general, using consent as a lawful basis for personal data processing is controversial. Consent is an expression of the paradigm of informational self-determination, designed to give data subjects 'control' over their personal data.[112] However, as underlined above, there is now broad empirical evidence questioning whether data subjects really are in a position to make such informed choices as they by and large do not understand the complexity of contemporary data flows.[113] Indeed, many individuals seem unaware of all the kinds of data processed by controllers, including what has been termed as 'bastard data': where the merging and comparing of data results in additional personal data.[114]

The challenge of acquiring informed consent for service improvements, specifically, lies in explaining the value of improved results vis-à-vis the various costs of collecting different pieces of user data. On the one hand, it is hard to expect the service provider to ask for such fine-grained consent when studies show that users have a limited understanding of the overall digital ecosystem, with some users not even aware that data such as search queries is stored and collected in the first place.[115] On the other hand, existing studies on related problems show that it is feasible to directly ask users for their privacy preferences when it comes to feature collection[116], or indirectly estimate how much users value their data in the context of specific tasks such as disease predictions.[117] At the same time, several lines of research, including explainable AI, or uncertainty and risk communication[118], aim at communicating the outputs of computational systems to end users in an understandable way. While those lines of work (on privacy preferences and outcome communication) are largely separate, it is feasible to imagine combining both approaches to design informed consent solutions for service improvements in personalisation, profiling, and decision-making systems.

Scholarship has long warned that consent 'should not bear, and should never have borne, the entire burden of protecting privacy'.[119] Consent is considered to be mainly theoretical and devoid of practical meaning, particularly since many Internet-based services cannot be used without consent.[120] It has indeed been suggested that the main feature of consent is 'to performatively legitimate otherwise unregulated unacceptable corporate practices.'[121]

There is thus broad scepticism regarding the suitability of consent as a legitimising basis. Furthermore, there is also reason to wonder whether the detailed requirements for valid consent can be met in the specific context of personalisation, profiling, and decision-making systems. The GDPR defines consent as 'any freely given, specific, informed and unambiguous indication of the data subject's wishes by which he or she, by a statement or by a clear affirmative action, signifies agreement to the processing of personal data relating to him or her'.[122] As a consequence, some forms of 'consent' such as pre-ticked boxes do not meet the GDPR threshold.[123]

In 2019, the French supervisory authority CNIL imposed a fine of €50 million on Google for having failed to get valid consent.[124] Important information such as the definition of the purpose was excessively disseminated across several pages, meaning that consent could neither be informed nor unambiguous or specific - the latter because users had to agree to the bulk of Google's terms before being able to use its services.[125] Consent appears to not be informed in most cases as a majority of users report not reading privacy policies of the services they use.[126]

Meeting the Regulation's requirements for valid consent is indeed extremely difficult in complex online data ecosystems. This can be seen in relation to real-time bidding, the process by which websites auction off personalised advertising space on websites in real-time.[127] The Interactive Advertising Bureau, a key industry organization, itself recognised that in real-time bidding, consent cannot be achieved as data subjects lack relevant information about data controllers.[128] Due to the complexity of such systems, data subjects are not in a position to understand the implications of clicking 'I agree'. In fact, research conducted in the United Kingdom in 2019 revealed that whereas 63% accept that online services are funded by advertisements, acceptance rates shift radically to only 36% once it is explained that personal data beyond browsing history is used to personalise ads.[129] This finding

110 Note that Article 8(2) of the Charter of Fundamental Rights refers to consent, and Treaty change would thus be necessary to remove consent as a valid ground for processing personal data.
111 Verbraucherzentrale Bundesverband, 'Zweckänderung in der EU-Datenschutzverordnung: Stellungnahme des Verbraucherzentrale Bundesverbands zum Expertengespräch zur Regelung der zweckändernden Weiterverarbeitung personenbezogener Daten in der EU-Datenschutz-Grundverordnung' (17 December 2014) https://www.vzbv.de/sites/default/files/downloads/EU-Datenschutzverordnung-BMI-Zweckaenderung-Stellungnahme-2014-12-17.pdf accessed 13 December 2019.
112 The European Commission has often underlined the GDPR's role in providing data subjects with control over personal data: European Commission, 'EU data protection rules' https://ec.europa.eu/info/priorities/justice-and-fundamental-rights/data-protection/2018-reform-eu-data-protection-rules/eu-data-protection-rules_en accessed 31 January 2020.
113 Omri Ben-Shahar and Carl Schneider, 'The Failure of Mandated Disclosure' (2011) 159 *University of Pennsylvania Law Review* 647.
114 'ENDitorial: Is "Privacy" Still Relevant in a World of Bastard Data?' (European Digital Rights (EDRi)) https://edri.org/our-work/enditorial-is-privacy-still-relevant-in-a-world-of-bastard-data accessed 3 December 2020
115 Catherine Miller, Rachel Coldicutt and Hannah Kitcher (n 36).
116 Andreas Krause and Eric Horvitz (n 12).
117 Gilie Gefen and others, 'Privacy, Altruism, and Experience: Estimating the Perceived Value of Internet Data for Medical Uses' [2020] *Companion Proceedings of the Web Conference 2020* 552.
118 David Spiegelhalter, 'Risk and Uncertainty Communication' (2017) 4 *Annual Review of Statistics and Its Application* 31.
119 Solon Barocas and Helen Nissenbaum, 'Big data's end run around procedural privacy protections', (2014) 57 *Communications of the ACM* 31, 33.
120 Bert-Jaap Koops (n 8) 251-252. , 'The Trouble with Data Protection Law' (2014) 4 International Data Privacy Law 250, 251-252.
121 Elettra Bietti, 'Consent as a Free Pass: Platform Power and the Limits of the Informational Turn' (2019) Pace Law Review (forthcoming).
122 GDPR, art 4(11) .
123 GDPR, recital 32 and Case C-673/14 Planet 49 [2019] ECLI:EU:C:2019:801.
124 Commission nationale de l'informatique et des libertés (CNIL), 'Délibération de la formation restreinte n° SAN – 2019-001 prononçant une sanction pécuniaire à l'encontre de la société GOOGLE LLC' (21 January 2019), SAN-2019-001.
125 It is worth noting that Google has appealed this decision.
126 Brooke Auxier and others, 'Americans and Privacy: Concerned, Confused and Feeling Lack of Control Over Their Personal Information' (Pew Research Center: Internet, Science & Tech, 15 November 2019) https://www.pewresearch.org/internet/2019/11/15/americans-and-privacy-concerned-confused-and-feeling-lack-of-control-over-their-personal-information accessed 3 December 2020.
127 For an overview, see Information Commissioner's Office, 'Update report into adtech and real time bidding' (20 June 2019) https://ico.org.uk/media/about-the-ico/documents/2615156/adtech-real-time-bidding-report-201906.pdf accessed 17 January 2019.
128 Johnny Ryan, 'New evidence to regulators: IAB documents reveal that it knew that real-time bidding would be "incompatible with consent under GDPR"' (*Brave*, 20 February 2019) https://brave.com/update-on-gdpr-complaint-rtb-ad-auctions accessed 17 January 2020.
129 Information Commissioner's Office, 'AdtechMarket Research Report' (March 2019) 5, 19 https://ico.org.uk/media/about-the-ico/documents/2614568/ico-ofcom-adtech-research-20190320.pdf accessed 17

indicates that, if consent really were informed, most users would not consent.

The requirement that consent be given 'freely' might also have far-reaching implications in personalisation. One may in fact wonder whether there is free consent in the absence of a non-personalised alternative. Recital 42 GDPR provides that there is no free consent 'if the data subject has no genuine or free choice or is unable to refuse or withdraw consent without detriment'.[130] Where a website cannot be used without consenting to personal data processing, 'the user does not have a real choice, thus the consent is not freely given'.[131] In 2013, the CJEU held that consent cannot be used as a lawful basis for fingerprinting in the process of obtaining a biometric passport as people need a passport and there is no alternative option available.[132] Although passports are arguably more essential than the use of specific online services, practical requirements to use the latter should not be underestimated (think, for instance, of the importance of search engines for contemporary lives or the significance of cloud computing providers for most businesses) and the Court's reasoning could also hold in relation to the latter. There thus appears to be a presumption that consent is invalid unless there is an alternative to use the service in a non-personalised way. As a consequence, consent 'should not generally be a precondition of signing up to a service'.[133] In 2018, an NGO brought a case (still pending) in Austrian courts that enquires whether consent is really free where users have no choice but to consent to continue using a service.[134]

What is more, pursuant to Article 7(4) GDPR, for consent to be freely given, 'utmost account shall be taken of whether, inter alia, the performance of a contract, including the provision of a service, is conditional on consent to the processing of personal data that is not necessary for the performance of that contract'. This indicates that there is a higher threshold for consent where it is used to justify a purpose that cannot be included in the initial purpose - itself governed by contract under Article 6(1)(b) GDPR.

For consent to be informed and to ensure transparency, 'data subjects/consumers should be given access to their 'profiles', as well as to the logic of the decision-making (algorithm) that led to the development of the profile'.[135] The requirement that data controllers disclose their 'decisional criteria' is considered particularly important as inferences can be more sensitive than the original data itself (a point we examine separately below).[136] This is an interesting statement as it may require a disclosure of the algorithm - contrary to what is generally considered necessary under Article 22 GDPR.

The feasibility of automatically checking for compliance with data processing declarations is another dimension pertinent to establishing whether consent mechanisms are meaningful. In the context of cookie processing, Santos et al. have argued for standardisation of consent in terms of both interfaces and language, as well as consent storage and withdrawal mechanisms.[137] Standardisation in this scenario might facilitate automated audits of data processing policies, enhance processing transparency, and increase the likelihood of consent being informed. The authors note, however, that in practice validating whether the consent policies are complied with will often require extensive manual validation. In data-driven systems, even if consent messaging as well as protocols were to be standardized, auditing for compliance would also require manual efforts. Crucially, these manual validations would have to be conducted in close cooperation with service providers. To the best of our knowledge, appropriate black-box auditing methods for compliance with service improvement purposes in data-driven systems have thus far not been developed.

A further practical question regarding consent as a legitimation of personalisation relates to Article 7(3) GDPR, which provides that data subjects can withdraw consent at any time. Whereas the withdrawal of consent does not negate the legitimacy of processing before withdrawal, it bars data controllers from continuing to process the data once the right has been revoked. This requirement would imply that, should a withdrawing user's data form a part of a trained model, the model might no longer be processed after consent is withdrawn.

It is far from established how to operationalise Article 7(3) GDPR in ML. Computers scientists have only recently started to develop solutions for efficient deletion of individual data points from trained machine learning models[138] and further research is necessary. At present, the complete removal of a user's data can often only be achieved by retraining the model from scratch on the remaining data, a procedure which is computationally costly and thus neither economical, practical or environmentally desirable.[139] It is worth noting that the same problem emerges where consent is exhausted once the purpose has been achieved.

Our analysis in this section has shown that the GDPR frames consent as a tool to get around purpose limitation requirements. Where an individual consents to expanded data processing, such processing can take place. This framing is problematic for a number of reasons. First, it minimises the effectiveness of purpose limitation. Second, it contributes to the increasing opacity of personal data processing (as examining purposes in terms of use rarely provides a transparent picture of what personal data is used for). Third, the specific legal requirements around consent—that it be freely given, specific, informed and unambiguous—can rarely if ever be meaningfully complied with. Yet, to date there has been insufficient enforcement of the legality of consent, as with the GDPR overall. Finally, there are currently no technical tools to efficiently implement the logical consequences of consent revocation.

January 2020.

130 GDPR, recital 42.

131 Eleni Kosta, 'Peeking into the cookie jar: the European approach towards the regulation of cookies' (2013) 21 *International Journal of Law and Information Technology* 380, 396.

132 Case C-291/12 Schwartz [2013] EU:C:2013:670, para 32.

133 See further Information Commissioner's Office, 'Guide to the General Data Protection Regulation (GDPR) – Consent' https://ico.org.uk/for-organisations/guide-to-data-protection/guide-to-the-general-data-protection-regulation-gdpr/lawful-basis-for-processing/consent accessed 18 October 2019; Article 29 Working Party, Guidelines on consent under Regulation 2016/679 (WP259 rev.01) 17/EN, 6; Frederik Borgesius et al, 'Tracking Walls, Take-It-Or-Leave-It Choices, the GDPR, and the ePrivacy Regulation' (2018) 3 *European Data Protection Law Review* 353, 361 (making this argument in relation to consent for tracking walls on websites).

134 Noyb, 'GDPR: noyb.eu filed four complaints over "forced consent" against Google, Instagram, WhatsApp and Facebook' (25 May 2018) https://noyb.eu/wp-content/uploads/2018/05/pa_forcedconsent_en.pdf accessed 17 January 2020.

135 Article 29 Working Party, Opinion 03/2013 on purpose limitation (WP 203) 00569/13/EN, 46.

136 Ibid, 47.

137 Cristiana Santos, Nataliia Bielova and Célestin Matte, 'Are Cookie Banners Indeed Compliant with the Law?' (2020) 2020 *Technology and Regulation* 91 https://techreg.org/index.php/techreg/article/view/43/25.

138 Antonio Ginart and others, 'Making AI Forget You: Data Deletion in Machine Learning' (2019) 32 *Advances in Neural Information Processing Systems* 3518.

139 Ibid.

4.4 Trade-Offs Inherent to Purpose Limitation

Purpose limitation comes with a number of considerable trade-offs. First, there is the explicit and significant trade-off between an unlimited and limited processing of personal data. The GDPR is a recent legislative affirmation of purpose limitation as a core tenet of data protection law. Data protection law ultimately serves to manage the risks that inevitably arise when personal data is processed and purpose limitation seeks to reduce such risks by limiting the ways in which the data can be processed. It has already been seen above that this limitation of data processing has in recent years been criticised as potentially stifling an innovative EU data economy, including in respect of artificial intelligence. It can be assumed that discussions about the desirability of purpose limitation will be revived in the coming years in light of envisaged legal reform (in the form of the proposed Data Governance Act and the expected AI Act) that would incentivise increased sharing and thus also repurposing of (personal) data. The promotion of data sharing services (also referred to as 'data marketplaces', essentially intermediaries that match data providers and data users) questions the very validity of purpose limitation. As such, we can expect an explicit and heated debate as to whether purpose limitation stands in the way of data sharing and the related expected societal benefits (such as in healthcare or climate change mitigation) in the EU.

Beyond this overarching explicit trade-off, our analysis has also revealed other trade-offs that were probably not envisaged by the legislative process. First, there is a trade-off between honesty and flexibility in purpose specification. It was observed that the purpose needs to be defined *ex ante*, yet any legitimate, sufficiently precise and explicit purpose meets the specification test. Data controllers might make a calculated decision as to whether to honestly define their present purpose or list different purposes not necessarily pursued in the present to cover potential future uses. Second, depending on the interpretation given to the research exemption, companies might have to make trade-offs in their organizational structures. If the exemption only applies to separate research teams, purpose limitation might disincentivise the creation of more integrated teams, even though such teams might be more beneficial in other respects.

4.5 Interim Conclusion

Our examination of the application of purpose limitation to personalisation, profiling and decision-making systems has revealed that purpose specification is a largely procedural criterion that does not really limit the ways in which personal data can be processed. While the compatible use requirement does aim at substantially limiting processing, there remains considerable uncertainty regarding the interpretation of the exemptions related to scientific research and statistics in data-driven systems. Furthermore, the limitations around data subject consent are not enforced in practice.

This does not, however, mean that the purpose limitation principle fulfils no function in data protection law. First, it forces controllers to ponder the need for and implications of personal data processing from the beginning. Second, respecting related requirements provides assurance to good-faith data controllers that processing is lawful. This echoes some elemental features of the GDPR, such as its role as a risk-management framework[140] (there is a recognition that processing generates risks and thus ought to be limited to what is necessary) and the balancing of the rights and interests of data subjects and controllers (in this case recognising that data controllers have an interest in processing data but their interests must be balanced against those of the data subject).[141] Nonetheless, our analysis above has shown that purpose limitation does not by itself stand in the way of profiling or personalisation systems. It does, however, result in numerous trade-offs, some of which might have been unintended. Below, we examine whether the same conclusion holds in relation to data minimisation.

5. Data Minimisation

Data minimisation is the logical consequence of purpose limitation. Article 5(1)(c) GDPR provides that data shall be 'adequate, relevant and limited to what is necessary in relation to the purposes for which they are processed'. It requires that no more personal data than necessary to achieve the purpose is processed and is also one of the 'technical and organizational measures' under Article 25(2), which reiterates that controllers only process personal data 'necessary for each specific purpose of the processing'. Thus, data minimisation should be engineered relative to the purposes.[142] Like purpose limitation, data minimisation is a risk-management measure as processing of excess data creates unnecessary risks from 'hacking to unreliable inferences resulting in incorrect, wrongful, and potentially dangerous decisions.'[143] Such risks can be minimised by making sure that controllers do not have more data than necessary and process it for no longer than necessary. Minimising the amount of data may even, depending on context, improve the quality of ML as there is less need to clean the data and less risk of inaccuracy (where the right data is chosen). Indeed, the quality of the training data and the features can be more determinative of model accuracy than the quantity of the training data.[144] To provide further context to these discussions, this section examines, from a legal and computational perspective, the three distinct components of data minimisation, namely that data must be (i) adequate, (ii) relevant, and (iii) limited to what is necessary in relation to the purposes for which they are processed.

5.1 Relevance

The GDPR requires that data processed for a given purpose be 'relevant'. Whereas this term has not been authoritatively defined, it appears to require that only pertinent data is processed.[145] Thus, a controller that processes irrelevant data breaches the principle. Imagine, for example, the scenario of an e-commerce website that requests your complete date of birth to provide personalised recommendations for future purchases. Unless its recommendations are supposed to have an astrological flavour, this data is irrelevant as indeed, it is likely that the company would be collecting this data to ends different from the stated purpose.

Seen from this perspective, relevance is designed to safeguard against the accumulation of data for the sake of gathering data or for undisclosed ends. There is no doubt that personal data has become

140 See further Recital 75 GDPR.

141 On the GDPR and risk management, see also Michèle Finck and Frank Pallas (n 30).

142 Sophie Stalla-Bourdillion and Alison Knight, 'Data Analytics and the GDPR: Friends or Foes? A Call for a Dynamic Approach to Data Protection Law' in Ronald Leenes et al (eds), *Data Protection and Privacy: The Internet of Bodies* (Hart 2018), 249.

143 Mireille Hildebrandt, 'Primitives of Legal Protection in the Era of Data-Driven Platforms', (2018) 2 *Georgetown Law Technology Review* 252, 267.

144 Datatilsynet, 'Artificial intelligence and privacy' (January 2018), 11 https://www.datatilsynet.no/globalassets/global/english/ai-and-privacy.pdf accessed 13 December 2019.

145 The French version of the GDPR indeed translates 'relevant' as 'pertinent'. The German language version of the GDPR indeed speaks of 'dem Zweck angemessen' - 'relevant for the purpose' in this context.

an extremely valuable commercial asset and there are incentives for controllers to accumulate a maximum thereof to develop their own business model, for speculative later use, or to re-sell.[146] It is worth noting however that, as stated by the International Working Group on Data Protection in Telecommunications, the capabilities that AI systems provide 'are pushing the limits for what is relevant, and the push to provide more and more data to facilitate connections pushes the data minimisation principle' as data becomes more meaningful when combined with 'other data, greater processing capacity and deeper analyses'.[147] Given the risks associated with an uncontrolled accumulation of personal data, the GDPR imposes limits on such practices. It moreover requires that personal data be adequate.

5.2 Adequacy

The requirements of relevance and adequacy are closely intertwined. Yet, there appears to be a nuance between both concepts. Whereas the relevance criterion has a purely limiting impact on data collection, in some circumstances adequacy may require that more data be processed. Indeed, omission of certain kinds of data can limit the usefulness and accuracy of a dataset and the analyses done on that dataset.[148] Minimisation is but one of various substantive requirements in Article 5 GDPR, others including fairness, transparency[149] and accuracy.[150] This provision ought to be interpreted holistically and its principles are to inform data minimisation and vice-versa. Using adequate data is indeed a means to ensure that a model is fair, transparent and accurate.

In some circumstances, adequacy will have a limiting effect on the quantity of data to be processed, such as where data that is inadequate in light of the purpose for which it is collected - as would be the case of the e-commerce website scenario above. However, in other circumstances, adequacy may require the processing of more data for data analysis to be fair and accurate. For example, it has been reported time and time again that many currently deployed models are inaccurate when it comes to certain demographic groups underrepresented in training datasets. In such instances, processing more data could make the corresponding model more representative and thus help achieve the overarching requirements of fairness, transparency and accuracy.

Although formulated as part of the data 'minimisation' requirement, it hence seems that the adequacy requirement can actually mandate the processing of more rather than less personal data. The final requirement of the test under Article 5(1)(c), necessity, in contrast has a purely limiting scope.

5.3 Necessity

Finally, data should be 'limited' to what is necessary, meaning that controllers ought to identify the minimum amount of personal data needed to fulfil a purpose.[151] This is a somewhat stricter wording compared to the DPD, which required that personal data must not be 'excessive in relation to the purposes.'[152] As a consequence, anything exceeding the 'minimum' amount necessary will be an excessive processing, in breach of the data minimisation principle. For example, where the same results can be achieved through the processing of less personal data, or even of anonymous data, the processing of personal data can likely not be accepted as necessary.

It is worth noting that where there are multiple purposes, a data item can be necessary for one purpose but not for another, and the data controller can only process for the former. The necessity criterion is also crucial for the interpretation of Article 7(4) GDPR which requires that when assessing whether consent is freely given, 'utmost account shall be taken of whether, inter alia, the performance of a contract, including the provision of a service, is conditional on consent to the processing of personal data that is not necessary for the performance of that contract'.

These findings confirm that data minimisation continues to play a meaningful function in contemporary data processing practices. First and foremost, its relevance and necessity requirements impose limits on the quantity of data that can be processed. One could argue that an ill-intended controller could define a purpose in such a manner that the data they want to collect is "relevant" (returning to the above example, they could state that they explicitly want to provide astrological recommendations). Yet, the necessity and adequacy imperatives impose limits on the boundless collection of personal data even in cases like this. What is more, adequacy ensures that the right kind of data is collected, also in furtherance of other GDPR objectives such as adequacy and fairness. Finally, data minimisation requires controllers to preferentially process personal data that constitutes less risk for data subjects.

5.4 What Data Needs to be Minimised?

Article 5(1)(c) GDPR is generally interpreted as referring to the need to minimise the *quantity* of data that is processed. One may, however, also wonder whether the principle extends to other characteristics such as whether the data has been pseudonymised or whether it is special category data. This includes data on racial or ethnic origin, political opinions, religious or philosophical beliefs, trade union membership, genetic data, biometric data used for the purpose of uniquely identifying a natural person, health data and data concerning a natural person's sex life or sexual orientation.[153] What data actually qualifies as sensitive data at a time where sensitive characteristics can often be inferred from behaviour traces is a matter of ongoing debate.[154]

First, personal data should be anonymised or pseudonymized wherever possible. Whereas perfect anonymisation, which is hard to achieve, brings the processing outside the scope of the GDPR altogether, pseudonymisation can reduce the risk inherent to the processing. This position also seems to find support in the E-Privacy Directive, which speaks of the need 'of minimising the processing of personal data and of using anonymous or pseudonymous data where possible'.[155]

Second, Article 9 GDPR establishes a special regime for categories of data considered to reveal particularly sensitive information about indi-

146 https://edpb.europa.eu/sites/edpb/files/files/file1/edpb_letter_out2020_0004_intveldalgorithms_en.pdf, 2-3.
147 International Working Group on Data Protection in Telecommunications, Working Paper on Privacy and Artificial Intelligence (n 6) 9.
148 Bart van der Sloot, 'From Data Minimization to Data Minimummization' in Bart Custers et al (eds) Discrimination and Privacy in the Information Society (Springer 2013) 274.
149 GDPR, art 5(1)(a).
150 GDPR, art 5(1)(d).
151 Information Commissioner's Office (n 68).
152 DPD, art 6(1)(c).
153 Article 9(1) GDPR.
154 See by way of example, Paul Quinn and Gianclaudio Malgieri, 'The Difficulty of Defining Sensitive Data – the Concept of Sensitive Data in the EU Data Protection Framework, *Brussels Privacy Hub Working Paper* (2020).
155 Directive 2002/58/EC of the European Parliament and of the Council of 12 July 2002 concerning the processing of personal data and the protection of privacy in the electronic communications sector (e-Privacy Directive) [2002] OJ L201/37, recital 9.

viduals. Article 9(1) establishes a general prohibition to process special category data (often also referred to as 'sensitive' data). In some circumstances, such data can nonetheless be processed where the data subject has provided explicit consent.[156] Under the GDPR, special category data can thus only be processed subject to conditions that are more burdensome for controllers than those arising under the ordinary regime. Explicit consent is the most relevant for profiling and personalisation systems. Whereas the concept of 'explicit consent' is not defined, it likely requires an oral or written affirmation of consent.[157] Sensitive data also ought not to be used to inform solely automated decisions that have legal or similarly significant effects on data subjects unless the data subject has explicitly consented or processing is necessary for reasons of substantial public interest.[158]

Personal data processed for profiling and personalisation often constitutes sensitive data,[159] such as when dating apps share their users' dating choices, information about drug use and ethnicity as well as precise geographical location with advertisers.[160] It is accordingly of pronounced practical importance for controllers of profiling and personalisation systems to determine whether their processing is caught by the GDPR's special regime.

Many agree that data minimisation not only entails an obligation to restrict the amount of data but also to keep sensitive data to a minimum. According to the Norwegian data protection authority, data minimisation 'stipulates proportionality' in intervening with a data subject's privacy.[161] This implies an obligation to restrict 'both the amount and the nature of the information used'.[162] As a consequence, pseudonymisation is encouraged as one measure limiting the identifiability of the data subject.[163] Zarsky concurs that minimisation 'relates to the scope and categories of data initially collected'.[164] This reflects that minimisation should not be seen as an isolated requirement but rather as a tool to interpret the entire GDPR. Indeed, the special regime created for special category data would substantiate the argument that the legislator intended for the processing of special category data to always be minimised.

Thus, data minimisation requires a limitation of sensitive data and at the same time, the latter is a frequent ingredient in personalisation systems. It is, however, doubtful that there is a legitimate basis for processing the data in light of the difficulty of achieving (explicit) consent. Supervisory authorities have for instance concluded that current consent requests in adtech do not comply with the requirements for explicit consent.[165] It thus appears that many profiling and personalisation systems are currently not compliant with the GDPR.

5.5 Current State

Despite the existence of appropriate computational techniques and empirical evidence suggesting that it might be possible to limit data in data-driven systems to a much larger extent, minimisation of user-generated, observational, and behavioral data does not appear to be a common practice. A study of software developers' approaches to data minimisation revealed that practitioner practices differ both in terms of protocols and tools,[166] highlighting the need for more specific implementation guidelines. The study, however, did not cover approaches to data minimisation in data-driven systems and, to the best of our knowledge, no such study exists.

Guidelines for implementing data minimisation have been issued by both the British[167] and Norwegain[168] data protection authorities. These guidelines suggest techniques that could be used to minimise data such as investigation of learning curves—a technique which ties minimisation to performance, similarly to one of the recently proposed formal minimisation interpretations.[169] Still, the suggestions do not go into lower-level operational details. Several open computational questions are a potential reason why more detailed guidelines for performance-based data minimisation are missing. The next section presentes these questions in detail.

5.6 Computational Challenges

Minimising data for the purpose of improving a system's performance faces a number of obstacles. The first and foremost challenge lies in determining whether and which data improves results. The state-of-the-art computing knowledge does not provide off-the-shelf answers. The closest relevant line of work aims at quantifying the impact of individual data points in training sets on the accuracy of machine learning models trained using those sets.[170] Such data points often correspond to individual persons and are composed of multiple pieces of information (features). To the best of our knowledge, methods that would quantify which individual pieces of data about an individual data subject are necessary to improve a personalized service (for the individual or globally) or quantify the improvement itself, are missing. The problem in fact poses a number of computational and research challenges.

Furthermore, determining whether a piece of personal data should be minimized out will be a form of prediction about future system performance. As such, these predictions can reasonably be expected to be inaccurate and a question remains what level of inaccuracy would be acceptable for this form of data minimisation to be deemed practically viable.

Beyond the prediction accuracy, the determination of how much and what personal data is to be kept for a given performance purpose, depends on a number of factors. Those include the prediction method itself, the underlying service algorithm, existing user data, as well as the entirety of other data at the disposal of the service provider. Out of those factors, two merit special attention. First, data minimisation outcomes will largely depend on the underlying algorithm. Advanced systems employing complex models that need to learn many parameters require enough data to function properly.

156 GDPR, art 9(2)(a).
157 Information Commissioner's Office (n 68).
158 GDPR, art 22(4).
159 Sandra Wachter, 'Affinity Profiling and Discrimination by Association in Online Behavioural Advertising' (2020) *Berkely Technology Law Journal* (forthcoming).
160 Natasha Singer and Aaron Krolik, 'Grindr and OkCupid Spread Personal Details, Study Says' New York Times (13 January 2020) https://www.nytimes.com/2020/01/13/business/grindr-apps-dating-data-tracking.html accessed 31 January 2020.
161 Datatilsynet (n 144).
162 Ibid.
163 Ibid.
164 Tal Zarsky, 'Incompatible: GDPR in the age of big data' (2017) 47 *Seton Hall Law Review* 995, 1009.
165 For an overview, see Information Commissioner's Office, (n 127).
166 Awanthika Senarath and Nalin Asanka Gamagedara Arachchilage, 'Understanding Software Developers' Approach towards Implementing Data Minimization' [2018] arXiv:1808.01479 [cs] http://arxiv.org/abs/1808.01479 accessed 2 December 2020.
167 Information Commissioner's Office (n 68).
168 Datatilsynet (n 144).
169 Asia J Biega and others (n 16).0
170 See, for example, Richard Chow and others, 'Differential Data Analysis for Recommender Systems' [2013] *Proceedings of the 7th ACM conference on Recommender systems* 323; Amirata Ghorbani and James Zou, 'Data Shapley: Equitable Valuation of Data for Machine Learning' (2019) 97 *Proceedings of the 36th International Conference on Machine Learning* 2242.

Examples include deep learning methods, whose recent reemergence has been made possible by the availability of vast image datasets[171], or natural language understanding methods that benefited from comprehensive Web corpora[172]. Thus, by developing and implementing state-of-the-art solutions, service providers might be able to collect more personal data. On the other hand, infusing vanilla models with domain-specific knowledge might allow for better performance with smaller models, and thus less need for data collection[173]. These observations further lead to a number of questions of economic nature. If less data can be collected with custom models, will companies who can afford an internal research unit be able to minimise data better? If bigger models grant a data processor the right to collect more data, will companies who can afford the costly infrastructure necessary to operate those models be allowed to collect disproportionately more data?

The second factor influencing data minimisation that is worth highlighting is the complex balance and interdependence of the data of different users and the system performance for those users. It might be tempting to think of the system data as static when considering which pieces of an individual's personal data to minimise out. However, the data that is minimized out for a given individual might constitute the system training data for other individuals. Thus, minimisation of data for a single user will also influence the performance of the system for other users. The need for a global, systemic approach that at the same time works for each individual separately, makes reasoning about data minimisation ever so complex and raises the question of whether we should acknowledge minimisation dependencies analogous to privacy dependencies.[174]

Last but not least, taking a user's perspective, it is important to recognise that different people might have different *personal purposes* when using a service in a seemingly same way. For instance, a user might generate a movie rating purely to give the provider the information needed for personalized movie recommendations, in which case the performance-based minimisation appears appropriate. Another user, however, might generate movie ratings for personal archiving purposes—to store a log of movies they have seen. In this case, the storage of all personal ratings appears appropriate. As a result, it might be necessary to not only model the purpose of data collection by the service provider, but also the *purpose of data generation* by the user.

The need to model user personal purposes leads to two challenges. First, it is rather difficult to infer user intent from their behavior. Recognising people's search intents from text queries, for instance, continues to be an active research problem in information retrieval.[175] Second, users might engage with technology products in originally unanticipated ways. Some people, for example, use email—developed primarily as a communication tool—as a task management and archiving system. Studies show that some users send themselves emails with todos, reminders, or files to archive.[176]

Inaccurate detection of user personal purposes might lead to both under- and over-minimisation of data, depending on the context. If personal purposes were recognized correctly, a system might adapt its minimisation strategies to a particular intent. For instance, if a user generated certain data points purely for archiving purposes, a system might store the data separately from all other user data and not use it for any other purpose, such as improving personalisation. Data generated for personalisation, on the other hand, might be minimised based on performance goals.

5.7 Trade-Offs Inherent to Data Minimisation

Just as purpose limitation, data minimisation presents numerous trade-offs. We again see the explicit and acknowledged trade-off between the risks and benefits of personal data usage. Data minimisation is essentially a risk-management tool which minimises risks by limiting the quantities and categories of data that can be lawfully processed.

The practical application of data minimisation, however, also results in numerous unexpected trade-offs. First, we have demonstrated that the determination of whether a piece of personal data should be minimised is a form of prediction about future system performance—which may be inaccurate. Controllers with inaccurate performance prediction algorithms might be rewarded by seemingly legitimate increased personal data collection. Second, data minimisation may drive the data controller to employ algorithms which are less robust to minimisation to be able to collect more data. In case such algorithms also offer worse performance, end users would end up penalised with both increased data collection and decreased satisfaction.

Finally, our analysis has revealed that data minimisation has collective rather than purely individual consequences. Minimisation of a user's data will impact system performance for other users and it will be important to understand the impact of individual subject's choices and preferences on the collective system dynamics.

6. Purpose Limitation and Data Minimisation Highlight Important Trade-Offs in Data Protection Law

Beyond the trade-offs inherent to purpose limitation and data minimisation, our research has further exemplified a number of other trade-offs inherent to data protection law at large.

6.1 The Generality of Legal Principles and The Need for Computationally Operational Interpretations

In order to comply with purpose limitation and data minimisation, computer scientists need measurable definitions of those principles as well as specific implementation guidelines. Only with precise mathematical definitions can algorithms determine which data to retain and which to discard, or predict whether data will improve services as it is collected. This need stands in tension with the GDPR as a general, principles-based and technology-neutral legal framework. Indeed, the GDPR and its implementing guidance do not provide any concrete indications to computing practitioners as to how to practi-

171 Jia Deng and others, 'ImageNet: A Large-Scale Hierarchical Image Database' [2009] 2009 *IEEE Conference on Computer Vision and Pattern Recognition* 248.

172 Alon Halevy, Peter Norvig and Fernando Pereira (n 11).

173 For instance, smaller custom-tailored models have been shown to outperform vanilla language models in dialogue systems. See: Matthew Henderson and others, 'ConveRT: Efficient and Accurate Conversational Representations from Transformers' [2020] *Findings of the Association for Computational Linguistics: EMNLP 2020* 2161.

174 Solon Barocas and Karen Levy, 'Privacy Dependencies' (2020) 95 *Washington Law Review* 555.

175 For example: Hamed Zamani and others, 'Generating Clarifying Questions for Information Retrieval' [2020] *Proceedings of The Web Conference 2020* 418; Bernard J Jansen, Danielle L Booth and Amanda Spink, 'Determining the User Intent of Web Search Engine Queries' [2007] *Proceedings of the 16th international conference on World Wide Web* 1149.

176 Horatiu Bota and others, 'Self-Es: The Role of Emails-to-Self in Personal Information Management' [2017] *Proceedings of the 2017 Conference on Conference Human Information Interaction and Retrieval* 205.

cally and concretely implement their legal requirements. As a result, currently it is difficult to determine whether a given computation adheres to the pre-defined purpose or whether collected data is adequate, relevant and necessary. Perhaps unsurprisingly, practitioners apply various, often inconsistent, approaches to minimisation.[177]

This trade-off between the value of general legal principles and the practical need for concrete interpretations could be addressed from both the legal and computational ends. On the legal side, it might be possible to develop more specific guidance. The European Data Protection Board would have to issue concrete overall guidance which might then be rendered more concrete when implemented at a firm level (requiring collaborations between technical and legal experts).

On the computational side, researchers could develop new technical implementation proposals which then could be evaluated by legal experts. Many algorithmic techniques that will likely be useful for automating data minimisation already exist (including, for instance, feature selection, outlier detection, analysis of learning curves, or active learning), even though they need to be adapted to adequate interpretations of data minimisation and purpose limitation. A recent line of work in computer science offers a glimpse of how we might attempt to interpret the principles in personalisation and profiling systems. Biega et al. proposed to interpret the purpose of data collection in data-driven systems as improvement in system performance metrics.[178] Shanmugam et al. proposed a framework for data minimisation based on algorithmic performance curves,[179] while Goldsteen et al. proposed a framework leveraging data anonymisation techniques.[180]

Our analysis has, however, highlighted the difficulties of automating legal compliance. Indeed, it may well be that in many scenarios measuring compliance with purpose limitation and data minimisation is simply too burdensome and costly. Similar difficulties can also be observed in respect of the computational implementation of another core GDPR principle, fairness.[181] Indeed, recent interdisciplinary work has highlighted that the legal prohibitions of certain kinds of discrimination (conventionally considered to be at the core of fairness) are too contextual, reliant on intuition and open to judicial interpretation to be automated. Thus, it is likely that many of the computational implementations of fairness, including "fairness toolkits" are unable to adequately reflect legal requirements.[182] Whereas this paper will not elaborate on these discussions in further detail, future work should more closely evaluate both the desirability and necessity of automating legal compliance. If automating compliance requires a fundamental change in law's contextual nature, discussions ought to be had about the implications and desirability of such changes.

The effort to translate the principles of purpose limitation and data minimisation into practice in data-driven systems will likely require an extensive dialogue between the legal and computational communities to determine which interpretations are viable both legally and computationally, much like the dialogue that has happened for the principles of antidiscrimination or fairness, and which has spun a large body of research in both communities.

6.2 The Unacknowledged Trade-Offs Between Various GDPR Principles

The above analysis has illustrated that the legal data minimisation principle requires that data usage is kept to the necessary minimum. Data minimisation hence encourages a restrictive processing of data, assuming that such restricted processing is preferred from a data protection perspective. It is important to acknowledge, however, that there is a trade-off between restrictive data usage and other GDPR objectives, such as fairness. Indeed, complying with fairness (which, it is important to point out, remains under-defined from a legal perspective) may require processing more data. Recent empirical studies in the domain of recommender systems have suggested that limiting data might have disparate consequences for individuals[183] and user groups,[184] while minimisation of sensitive features (such as gender) may moreover limit our ability to audit fairness.[185]

Furthermore, Article 5 (1)(d) GDPR requires that personal data be "accurate and, where necessary, kept up to date; every reasonable step must be taken to ensure that personal data that are inaccurate, having regard to the purposes for which they are processed, are erased or rectified without delay". This accuracy requirement may also compel the data controller to collect more data. For instance, personalisation profiles consisting of product ratings or search queries may become obsolete if data collection stops because of the data minimisation requirement, yet the interests and preferences of data subjects change. In this context, it might seem necessary to continuously collect new data while focusing minimisation on the old data. In fact, Google introduced such 3-18 months auto-deletion of search and location data in 2019,[186] and made it the default setting for new users in 2020.[187]

The fact that data minimisation promotes reliance on restrictive quantities of data whereas other GDPR principles such as fairness and accuracy will sometimes require the collection of additional data raises the question of how these objectives ought to be reconciled in practice. All these requirements constitute core data protection principles enshrined in Article 5 GDPR—which does not establish a hierarchy among its various requirements. As such, it cannot be concluded that data minimisation is a superior objective compared to fairness or vice versa. Thus, in practice, computer scientists must make sure that data minimisation as well as fairness and accuracy are equally respected.

Data minimisation by itself already incorporates leeway for such balancing of principles through its three requirements of relevance, adequacy and necessity. The collection of further data to comply with the fairness or accuracy requirements can, depending on the circumstances, be considered to be relevant, adequate and necessary. Indeed, our analysis above confirmed that the adequacy requirement can itself be read as requiring the collection of more data to comply with considerations such as fairness. From a legal perspective, however, what is relevant, adequate and necessary should be determined

177 Awanthika Senarath and Nalin Asanka Gamagedara Arachchilage (n 166).
178 Asia J Biega and others (n 16).
179 Divya Shanmugam and others (n 15).
180 Abigail Goldsteen and others (n 27)
181 Article 5(1)(a) GDPR.
182 Sandra Wachter, Brent Mittelstadt and Chris Russell, 'Why Fairness Cannot Be Automated: Bridging the Gap Between EU Non-Discrimination Law and AI' [2020] SSRN Electronic Journal https://www.ssrn.com/abstract=3547922 accessed 3 December 2020.
183 Asia J Biega and others (n 16).
184 Hongyi Wen and others (n 16).
185 Gemma Galdon Clavell and others, 'Auditing Algorithms: On Lessons Learned and the Risks of Data Minimization' [2020] *Proceedings of the AAAI/ACM Conference on AI, Ethics, and Society* 265.
186 Google 'Introducing Auto-Delete Controls for Your Location History and Activity Data' (*Google Blog*, 1 May 2019) https://blog.google/technology/safety-security/automatically-delete-data accessed 3 December 2020.
187 Google (n 17).

on a case-by-case basis, taking into account contextual factors.

6.3 Data Subject Rights and the Economic and Environmental Costs of Enforcing Them

Our analysis has highlighted that personal data can be processed for purposes exceeding the initially defined purposes with a data subject's consent. At the same time, the GDPR provides that whenever personal data processing is legitimised through consent, the data subject subsequently has the right to withdraw his or her consent at any time.[188] Whereas the withdrawal of consent does not affect the lawfulness of past personal data processing, it prohibits the data controller from continuing to process that personal data in the future. Machine learning models are trained on already collected personal data but are employed to make new inferences—a form of personal data processing. Thus, withdrawal of consent to process personal data encoded in a model requires deconstructing the model. How to efficiently remove individual data points from trained machine learning models is a subject of active research.[189] Currently, in cases of consent withdrawal, models might have to be continually retrained, yielding computational and thus also environmental costs. How to balance enforcement of individual data rights *vis-à-vis* environmental costs is a pertinent question.

6.4 The Cost of Compliance and the Unlikelihood of Enforcement

Ultimately, the success of policies, including data protection, hinges on their practical implementation. Unless there is adequate enforcement of related provisions, it is doubtful whether addresses have sufficient incentives to enforce related legal requirements, particularly if such implementation is costly. Compliance with data protection law is indeed costly. It requires data controllers to contract related expertise as well as carefully designing their technical and organizational structures. Perhaps most significantly, it also prevents them from pursuing forms of data analysis that may be attractive from a business perspective yet risky in terms of violating data protection law.[190] As such, it is important that data protection law is properly implemented. At present data controllers will rationally make a trade-off between the economic benefits of unconstrained usage of personal data and the potential yet very unlikely economic cost of data protection enforcement.

Recent years have, however, underlined that the enforcement of the GDPR is riddled with hurdles, such as the uneven geographical distribution of relevant competence (on the basis of a company's seat in the EU) or the fact that data protection authorities have insufficient means to meaningfully police compliance with the Regulation.[191] Indeed, even though Article 52(4) GDPR requires that supervisory authorities have the required technical resources, evidence is mounting that these resources are currently insufficient.[192]

Another factor that is highly relevant with respect to purpose limitation and data minimisation relates to the technical difficulties of verifying compliance. Indeed, whereas anyone, including supervisory authorities, can in most circumstances consult a data controller's data protection policies to read or automatically analyse[193] how the purposes are defined and what data is acknowledged to be processed, verifying whether these statements are honored in practice is an entirely different matter. Determining whether individual pieces of data are necessary for personalisation might be computationally difficult, and in general more research would be necessary to establish what is the form of evidence that data controllers should produce to prove compliance. Furthermore, measures of minimisation could be gamed. Since improvements in the results are often functions of complex interactions between individual pieces of data, it is feasible to imagine a data collection mechanism that requests a large set of data, where seemingly all items are necessary, even though there exists a smaller set of data yielding a similar performance that could have been collected instead. While continual reassessment of whether existing data is necessary is mandated, it might be impossible to determine whether the retained data in fact is the minimum data.

7. Outlook

This paper has shown that, despite what has been suggested by many commentators, purpose limitation and data minimisation remain feasible albeit challenging in the context of data-driven personalisation, profiling and decision-making systems. At the same time, they force data controllers to make many, oftentimes unacknowledged, trade-offs. While the longer-term research problems await their solutions, practitioners might employ a variety of organizational and technical best practices as well as off-the-shelf tools that minimise data even if not explicitly developed for minimisation purposes.

7.1 Short-term Practitioner Guidelines

Even though the implementation of purpose limitation and data minimisation in the context of data-driven systems bears a considerable research agenda, practitioners might consider implementing a range of existing solutions and best practices that contribute toward data minimisation. The first organizational best practice is for employers and employees to create and cultivate a mindset of reflecting on the purposes of data they collect, continuously considering whether data should be collected and when it should be deleted. To quantify the importance of different pieces of data, practitioners can use off-the-shelf solutions for machine learning models, including feature selection, data influence estimation, or data valuation. At the data collection time, techniques such as active learning would allow data processors to prioritize which data is the most important for a model's quality. Data can also be minimised through simpler heuristics, such as selection of representative random samples, or selection of data specifying certain domain-specific quality criteria, or retaining of the most recent data only. For instance, in the context of product recommendations, a data controller might retain only the most recent product ratings generated by a user, only the ratings for the

188 Article 7(3) GDPR.

189 Some of the recently proposed approaches include: Antonio Ginart and others (n 138); Lucas Bourtoule and others, 'Machine Unlearning' [2020] arXiv:1912.03817 [cs] http://arxiv.org/abs/1912.03817 accessed 3 December 2020; Sanjam Garg, Shafi Goldwasser and Prashant Nalini Vasudevan, 'Formalizing Data Deletion in the Context of the Right to Be Forgotten' (2020) 12106 *Advances in Cryptology – EUROCRYPT 2020* 373; Chuan Guo and others, 'Certified Data Removal from Machine Learning Models' (2020) 119 *Proceedings of the 37th International Conference on Machine Learning* 3832.

190 See also Articles 5(2) and 25(1) GDPR.

191 Derek Scally, 'German Regulator Says Irish Data Protection Commission Is Being "Overwhelmed"' (*The Irish Times*, 3 Feb 2020) https://www.irishtimes.com/business/financial-services/german-regulator-says-irish-data-protection-commission-is-being-overwhelmed-1.4159494 accessed 3 December 2020.

192 https://brave.com/wp-content/uploads/2020/04/Brave-2020-DPA-Report.pdf

193 Various natural language processing (NLP) techniques have been proposed to automatically extract or align policy statements; see, eg Shomir Wilson and others (n 79). Further techniques include automated question answering, allowing readers to obtain concise answers to their questions about a given verbose policy, see Abhilasha Ravichander and others (n 79).

most popular products, or only the ratings with the highest or lowest values. Finally, minimisation should be employed not only at the data collection time, but also continually reapplied to existing data stores.

7.2 Long-term Research

To be technically implementable in the context of data-driven systems, purpose limitation and data minimisation will likely need to follow a similar research trajectory as that of work in algorithmic fairness. As demonstrated throughout the paper, we need mathematical interpretations of the principles, decision rules for deciding which pieces of data are necessary and which should be discarded, machine learning models that could automate compliance, and quantitative data analyses for understanding how the implementation of those principles might influence the quality and functioning of online ecosystems. We need standardisation of data processing purposes to ensure their specificity as well as an understanding of how different purposes relate to each other to reason about their compatibility. We lack an in-depth understanding of what value people associate with different types of data in different contexts. We moreover should design appropriate transparent mechanisms for collecting informed data processing consent as well as technical means of removing data from existing models and infrastructures once a purpose has been fulfilled or a user withdraws their consent. We need auditing methods that could establish compliance with the purpose limitation and data minimisation requirements. Finally, we need to establish normatively, legally, and technically, how to balance data minimisation with other GDPR requirements—such as fairness or accuracy—which might be at odds with the minimisation principle.

Yet, as the work on algorithmic fairness has previously exemplified, it is difficult to bridge terminological and substantive gaps between disciplines.[194] One may wonder whether and how legal principles, particularly broad principles purposefully kept vague to enable contextual interpretation, can be translated into computer code. Without doubt, this is a question at the heart of digitalisation that requires more engagement from multiple disciplines in the years to come.

Acknowledgements

We thank Solon Barocas for discussions that helped ideate the interdisciplinary direction of this work, as well as Hal Daumé III, Fernando Diaz, Peter Potash, Samira Shabanian, and Divya Shanmugam for fruitful discussions on computational interpretations of purpose limitation and data minimisation. We are also grateful to Kai Ebert for excellent research assistance.

194 Sandra Wachter, Brent Mittelstadt and Chris Russell (n 182); Ben Green and Salomé Viljoen, 'Algorithmic Realism: Expanding the Boundaries of Algorithmic Thought' [2020] *Proceedings of the 2020 Conference on Fairness, Accountability, and Transparency* 19.

05

Technology and Regulation

The Right of Access to Personal Data: a Genealogy

René Mahieu*

Right of Access to Personal Data, Due Process, Informational Self-Determination, Power Reversal, Information Asymmetry, Civil Rights, GDPR

rene.mahieu@vub.be

In this paper, I analyze several traditions of data protection to uncover the theoretical justification they provide for the right of access to personal data. I find little support for the claim that the right follows from the German tradition of "informational self-determination" or Westin's idea of "privacy as control". Instead, two other less known theories of data protection appear to offer a direct justification for the right of access. First, Westin and Baker's "due process" view, which access helps to expose error and bias in decision-making, thereby contributing to correct decisions and allowing affected people to be involved in the decision making. Second, Rodotà's "power reversal" view of access which enables social control over the processing of personal data and serves as a counterbalance to the centers of power by placing them under the control of democratic accountability.

"Ours is a society that has always expected law to define basic citizen rights, and the scope of what American society regards as rights and not privileges has been widened dramatically in the past decade."

– Alan Westin and Michael Baker, *Databanks in a Free Society*[1]

"The regulation of the right of access imposes a completely new regulation of secrecy and opens up the possibility of new developments in civil rights, expanding the knowledge available to citizens, and thus their power of control over public and private action."

– Stefano Rodotà, *Computers and Social Control*[2]

1. Introduction

The foundations of data protection law were laid down in the 1960s and 1970s. Directly from the start, the right of access to personal data was included in all major data protection regimes in the world,[3] and became a cornerstone of data protection legislation,[4] which it remains until today. The relevance of the right of access is evidenced – among other things – by the fact that it is part of the Charter of Fundamental Rights of the European Union (The Charter).[5] Moreover, its continued importance is confirmed by the fact that strengthening the right of access and other data subject rights was one of the core objectives of the introduction of the European General Data Protection Regulation (GDPR).[6] Yet, while the importance of the right of access to personal data is generally assumed, there is no comprehensive and detailed account in recent literature of why this right is so important and what purposes it is supposed to serve.

Against this background, this article presents an investigation into

1 Alan F Westin and Michael A Baker, *Databanks in a Free Society* (Quadrangle Books 1972) 347.

2 Stefano Rodotà, *Elaboratori Elettronici E Controllo Sociale [Computers and Social Control]* (Societa Editrice Il Mulino 1973) 67. No English translation of Elaboratori Elettronici E Controllo Sociale has been published. All translations are by the author, two Italian native speakers specialized in data protection law – Ilaria Buri and Simone Casiraghi –, and with the help of translation service https://www.deepl.com.

3 Colin J Bennett, *Regulating Privacy: Data Protection and Public Policy in Europe and the United States* (1st edn, Cornell University Press 1992) 106.

4 In this article I will use the term "data protection", which is the common terminology in Europe for something that is similar to what in the US is generally called "data privacy". While there is much ado about the differences between these concepts, and the differences between EU and US regulation (See for example generally: Anupam Chander, Margot E Kaminski and William McGeveran, 'Catalyzing Privacy Law' [forthcoming] *Minnesota Law Review* https://www.ssrn.com/abstract=343392> accessed 25 January 2020), these differences seem to have been much less pronounced in the early days of data protection.

* René Mahieu is a PhD candidate at the Law, Science, Technology & Society research group (LSTS) at VUB Brussels.

Received 17 Apr 2021, Accepted 12 Aug 2021, Published 20 Aug 2021.

5 Charter of Fundamental Rights of the European Union, OJ 2010 C 83/389. The relevant Art. 8(2) of the Charter of Fundamental Rights of the European Union reads: 'Such data must be processed fairly for specified purposes and on the basis of the consent of the person concerned or some other legitimate basis laid down by law. *Everyone has the right of access to data which has been collected concerning him or her, and the right to have it rectified' [Emphasis added].*

6 European Commission, COM(2010) 609 final COMMUNICATION FROM THE COMMISSION TO THE EUROPEAN PARLIAMENT, THE COUNCIL, THE ECONOMIC AND SOCIAL COMMITTEE AND THE COMMITTEE OF THE REGIONS - A comprehensive approach on personal data protection in the European Union 7-8 (2010), https://eur-lex.europa.eu/legal-content/EN/TXT/PDF/?uri=CELEX:52010DC0609&from=EN; Viviane Reding, 'The European Data Protection Framework for the Twenty-First Century' (2012) 2 *International Data Privacy Law* 119, 124-126.

the politico-philosophical origins of the right of access to personal data. To find justifications for the right of access to personal data, an extensive review of the literature has been conducted. This review included (1) literature on the value of data protection and data subject rights, (2) studies on the right of access, (3) legislative history (on the level of the EU, the Netherlands, Germany, France, and International Organizations, such as the OECD and Council of Europe). Through this broad review, four perspectives were identified, two of which (1) provide a detailed account of the value and purpose of the right of access to personal data, and (2) were developed by scholars that had a considerable influence on European data protection law.

Many academic accounts that consider the justification for data subject rights argue or assume that it belongs to either "informational self-determination" and/or "privacy as control".[7] Norris and L'Hoiry, for example, write that "Access to personal data is the natural pre-condition of data subjects' ability to exercise the remainder of their ARCO rights (access, rectification, cancellation, opposition). Put simply, citizens cannot exercise their rights of *informational self-determination* in an informed manner without knowing what is held about them."[8] In European policy making too, data subject rights are often understood within a narrative of control over data.[9]

However, the relationships between those theories and the right of access are rarely formulated in depth, nor fully convincing. One of the reasons for this is probably that access rights were not a central element in these doctrines of data protection, as a closer look at the historical roots of those theories will show.

Textual interpretation of the laws, even when analyzed in conjunction with their accompanying legislative materials (i.e. policy documents and legislative histories), often provide only a limited view of the functions they fulfill. As other scholars have already noted, this is certainly the case with regards to data protection laws.[10] There are several reasons for this lack of clarity. First, laws are being made in complex institutional structures and are often the result of political compromise, edging off the clarity of the initial ideas which pave the way for the introduction of these laws. Furthermore, the mere fact that the right of access to personal data has already been part of the data protection regimes for such a long time makes it likely for it to be part of any new law without much renewed discussion of the principles and ideas on which it is based.[11]

Two other data protection theories exist in which the right of access did have a central role, and which also had a very direct influence on the development of European data protection legislation. In the United States, Alan Westin, together with Michael Baker developed the view that the constitutional principle of due process should apply to the processing of personal data. In this view, the right of access to personal data is essential for the protection of due process. At the same time, Italian scholar Stefano Rodotà – following a tradition of critical legal theory – developed the view that access to personal data should serve as a general counterbalance to the power asymmetries associated with the accumulation of data. Both gave detailed accounts of the importance of the right of access and placed it at the center of their proposals for data protection regulation. However, while their work had a significant influence on the development of data protection regulation, it has remained broadly overlooked in contemporary debates on data protection and the right of access.

In order to find the politico-philosophical origins and justification of the right of access to personal data, I discuss the four above-mentioned theories of data protection. First (in Section 2), I argue that, for Westin, access rights are not intended to safeguard "privacy as control". Instead, I show how Westin, in his most famous book *Privacy and Freedom*[12], starts to develop the idea that people should have a right to access to data to protect their "due process" rights in an age of electronic data processing, and how he later develops this theory fully, together with Michael Baker, in *Databanks in a Free Society*. Then (in Section 3) I discuss the theory of informational self-determination and show that the right of access does not have a central position in that theory. In Section 4, I discuss Rodotà's "power reversal" view of data protection, in which the right of access – and in particular the collective use of that right – plays an essential role.

In Section 5, I discuss the wider implications of the preceding analysis. In particular, I highlight how the historical analysis shows that access rights are conceptualized as a way to empower people in relations characterized by structural informational power asymmetry. And that while the right enables people to gain access to data (or, in the language of Westin and Baker: "files"), the ultimate aim of the right is

7 See also Antoinette Rouvroy and Yves Poullet, 'The Right to Informational Self-Determination and the Value of Self-Development: Reassessing the Importance of Privacy for Democracy' in Serge Gutwirth and others (eds), Reinventing Data Protection? (2009) 69 https://doi.org/10.1007/978-1-4020-9498-9_2; Jef Ausloos, *The Right to Erasure: Safeguard For Informational Self- Determination In a Digital Society?* (Doctoral Thesis, KU Leuven 2018) section 2.2.2 and 2.3.3; HU Vrabec, *Uncontrollable: Data Subject Rights and the Data-Driven Economy* (Doctoral Thesis, Leiden University 2019) chapter 4 https://openaccess.leidenuniv.nl/handle/1887/68574. Bart Van der Sloot, 'Do Data Protection Rules Protect the Individual and Should They? An Assessment of the Proposed General Data Protection Regulation' (2014) 4 *International Data Privacy Law* 307 http://dx.doi.org/10.1093/idpl/ipu014; ; Orla Lynskey, *The Foundations of EU Data Protection Law* (Oxford University Press 2015).
Most of the authors that relate the right of access to informational self-determination do not make very specific assertions about the nature of this relationship.
Rouvroy and Poullet (2009, 69), for example, remain quite general and state that all the main principles of data protection "might be viewed as a development of the self-determination principle in the area of the personal data flows" and claim that the purpose of access rights is "allowing a better control over the uses and dissemination of personal data". Van der Sloot (2014), for example, who himself does not subscribe to this position, claims that other scholars relate the right of access and control by the individual data subject to informational self-determination he does not point at any specific scholars who do so.
According to Lynskey (Chapter 6) the data subject rights, including the right of access which she sees as "the foundational block on which other rights of control rest" in European data protection law are there to enebale individuals to exercize individual control over personal data, and she relates this to the conception of privacy-as-control as described by Westin. Similarly, Ausloos (p. 73-74) writes that the data subject rights are the material implementation rationale of data subject control, in Westin's sense, in data protection law.

8 Clive Norris and Xavier L'Hoiry, 'Exercising Citizen Rights Under Surveillance Regimes in Europe – Meta-Analysis of a Ten Country Study' in Clive Norris and others (eds), The Unaccountable State of Surveillance: Exercising Access Rights in Europe (Springer International Publishing 2017) 405 https://doi.org/10.1007/978-3-319-47573-8_14.

9 E.g. European Commission (n 6).

10 E.g. Orla Lynskey, 'Deconstructing Data Protection: The "Added-Value" of a Right to Data Protection in the EU Legal Order' (2014) 63 *International and Comparative Law Quarterly* 569, 562. ("However, what is apparent from this scholarly speculation is that the EU has not adequately justified the introduction of the right to data protection in the EU legal order or explained its content."). https://www.cambridge.org/core/product/identifier/S0020589314000244/type/journal_article; https://core.ac.uk/download/pdf/191099366.pdf (open access).

11 Alexander Dix, 'Artikel 15 Auskunftrecht Der Betroffenen Person' in Spiros Simitis, Gerrit Hornung and Indra Spiecker, *Datenschutzrecht* (1st edn, Nomos Verlagsgesellschaft 2019), 651.

12 Alan F Westin, *Privacy and Freedom* (first published 1967, Ig Publishing 2015).

to enable people to understand and contest individual decisions, and even systems of decision-making which are based on personal data. This is relevant for a variety of current debates, such as the questions about the scope of access rights, the extent to which the right entails a "right of explanation",[13] or the collective aspects of that right.[14] Moreover, I discuss the right's emancipatory aim, which can only bear fruit once it is properly recognized. Finally, I reflect on how this historical analysis can contribute to the ongoing discussion on the values that are safeguarded by data protection more broadly.[15]

2. Westin and Baker: From Privacy as Control to Access as Due Process

In *Privacy and Freedom*, published in 1967, Alan F. Westin defined his famous notion of "privacy as control" as "the claim of individuals, groups, or institutions to determine for themselves when, how and to what extent information about them is communicated to others".[16] This idea of "privacy as control" is seen as a fundamental theory of data protection and privacy.[17] Moreover, his work was a key point of reference for the European data protection community, and had significant influence on the development of European data protection law.[18]

According to some authors, the right of access is part of this conceptualization of privacy as individual control over the flow of personal data.[19] In *Privacy and Freedom*, Westin does indeed mention, albeit very briefly, the right of access in relation to "privacy as control".[20] However, more importantly, he also introduces the idea that due process rights should apply to the processing of personal information. The second part of this section discusses *Databanks in a Free Society*, where the idea of applying due process to the processing of personal information is thoroughly developed and discussed in detail, and independently from the principle of privacy.

The wide array of privacy-related questions Westin deals with in *Privacy and Freedom* are mostly focused on surveillance made possible by new technological methods, such as wiretapping, subliminal suggestion, lie-detecting and personality testing. Data processing by electronic means, which has most resonance with what we currently call data protection, is but one of the many elements discussed in one chapter of this book.[21] Within this chapter on processing of data by computers, the elaboration of the right of access to personal data is laid out in just two pages and presented only in embryonic form. Westin's discussion of the right of access in this chapter should therefore be understood more as an initial thought experiment than as a full-fledged theoretical exposition.

Westin writes, towards the end of *Privacy and Freedom*, that "personal information, thought of as the right of decision over one's private personality, should be defined as a property right, with all the restraints on interference by private and public authorities and due-process guarantees that our law of property has been so skillful in designing".[22] In his view, due process as applied to the processing of personal data would include (1) a right to notice when information is put into a file; (2) a right to examine [access] the file; (3) a right to challenge accuracy; (4) a right to have the challenge recorded; and (5) a right to deletion in some cases.[23]

At a glance, the proposed rights could be seen as an extension of "privacy as control", as they follow from making a link between personal data and private property, and property relations conventionally being seen as the epitome legal form for allowing people to exert control. However, this reading would overlook the reason that he defines personal information as property, which is that "so defined, a citizen would be entitled to have due process of law". Thus, due process is not just a mere beneficial side effect of granting property rights. Instead, the fact that due process rights are connected to property – at least in the US Constitution – would be the primary reason for classifying personal data as private property. Moreover, Westin argues that assigning property rights to personal data would bring personal data under a whole range of additional strong legal protections that the US legal system affords to private ownership. It should be stressed that Westin's aim is to provide more protections to people with respect to their personal data, not to create a market for personal data.

There is an important secondary motivation for implementing these rights (now called data subject rights), which is completely unrelated to the logic of "privacy as control". Westin explains it as follows: "When the information keeper knows that the individual will be notified, can see and can challenge the information, all the restraints of visibility of action will be on the keeper. His loss of anonymity will

13 See generally e.g. Andrew D Selbst and Julia Powles, 'Meaningful Information and the Right to Explanation' (2017) 7 *International Data Privacy Law* 233 https://doi.org/10.1093/idpl/ipx022; Margot E Kaminski, 'The Right to Explanation, Explained' (2019) 34 *Berkeley Technology Law Journal* 189 https://btlj.org/data/articles2019/34_1/05_Kaminski_Web.pdf.

14 See generally, e.g. René LP Mahieu, Hadi Asghari and Michel JG Van Eeten, 'Collectively Exercising the Right of Access: Individual Effort, Societal Effect' (2018) 7 *Internet Policy Review* 1 https://doi.org/10.14763/2018.3.927; René LP Mahieu and Jef Ausloos, 'Harnessing the Collective Potential of GDPR Access Rights: Towards an Ecology of Transparency' [2020] *Internet Policy Review* https://policyreview.info/articles/news/harnessing-collective-potential-gdpr-access-rights-towards-ecology-transparency/1487.

15 See generally, e.g. Julie E Cohen, 'Turning Privacy Inside Out' (2019) 20 Theoretical Inquiries in Law 1 https://din-online.info/pdf/th20-1-3.pdf; Orla Lynskey, 'Delivering Data Protection: The Next Chapter' (2020) 21 German Law Journal 80 https://doi.org/10.1017/glj.2019.100.

16 Westin (n 12) 5.

17 Colin J Bennett and Charles D Raab, *The Governance of Privacy : Policy Instruments in Global Perspective* (MIT Press 2006).; Seda Gürses, 'Can You Engineer Privacy?' (2014) 57 *Communications of the ACM* 20, 21 https://cacm.acm.org/magazines/2014/8/177015-can-you-engineer-privacy/.

18 Work by Westin, and particularly *Privacy and Freedom*, is a primary reference for example in the development of the first German and Dutch data protection laws (See respectively: Ruprecht B Kamlah, 'Datenschutz Im Spiegel Der Angloamerikanischen Literatur -- Ein Überblick Über Vorschläge Zur Datenschutzgesetzgebung -- Report for the Ministry of the Interior' (1971) Drucksache VI/3826 Deutscher Bundestag — 6. Wahlperiode. Thijmen Koopmans (ed), *Privacy en persoonsregistratie: interimrapport van de Staatscommissie bescherming persoonlijke levenssfeer in verband met persoonsregistraties* (Staatsuitgeverij 1974), 6-7). Similarly, it heavily influenced the work on the right to informational self-determination as well as that by Stefano Rodotà (See respectively for informational self-determination: Mallmann Christoph Mallmann, *Datenschutz in Verwaltungs-Informationssystemen* (Oldenburg 1976), and for Rodotà: Rodotà (n 2) and section 4 below.

19 Bennett and Raab (n 18), 98-99; Antoinette Rouvroy and Yves Poullet (n 7) 45, 62 and 68-75 https://doi.org/10.1007/978-1-4020-9498-9_2. Bennett and Raab discuss the right of access to personal data, together with informed consent, as the primary privacy principles meant to empower individuals. Rouvroy and Poullet elaborate on the link between informational self-determination, privacy as control and privacy as empowerment. See also Ausloos (n 7); Lynsky (n 7).

20 Westin (n 12) 362 "First, personal information, thought of as the right of decision over one's private personality, should be defined as a property right, with all the restraints on interference by public or private authorities and due-process guarantees that our law of property has been so skillful in devising."

21 Westin (n 12) Chapter 12 "Pulling all the facts together".

22 Westin (n 12) 362.

23 Westin (n 12) 363.

be the best guarantee of fairness and care in the information-keeping procedure".[24] In other words, the right of access, by shedding light on the actions of the data controller, functions as a safeguard against the misuse of this data by the controller.

The theory of applying due process to the processing of personal data is only fully developed in *Databanks in a Free Society*, published by Westin and Baker in 1972, five years after *Privacy and Freedom*.[25] Here, the value of due process is completely independent from the value of "privacy as control", and rid of its connotations of private property. The right of the citizen "to see his record"[26] is no longer one of many policy proposals, but the main focal point.[27] In this book the need for, and purpose of, a right of access to personal data is elaborated in much more detail.

The research for *Databanks in a Free Society* was performed by the "Computer Science and Engineering Board of the United States National Academy of the Sciences", directed by Westin, who was at the time a professor of public law and government at Columbia University.[28] This academic work, as the name suggests, describes the effects that developments in electronic computing and the creation of new databanks have on the foundations of a free and democratic society. While the concluding analysis of the report is theoretical, the study is grounded on empirical and interdisciplinary research concerning the consequences of the introduction of electronic databases across society. Westin and his team visited the sites of databases, conducted in depth interviews with personnel on site and sent questionnaires. The main purpose of doing this study was to find how the introduction of computers affected the creation, sharing and use of files on individuals, in particular in relation to their civil liberties.[29]

Westin and Baker note that there are two fundamental constitutional principles that govern the processing of personal data: (1) privacy and (2) due process.[30] Due process is a doctrine of procedural safeguards that comprises a set of rights for citizens and obligations for the government with regards to decisions that affect citizens.[31] The overall purpose of these rules is to put a check on the arbitrary exercise of power by the state. The right to privacy, as "the right to be left alone", on the other hand is the right to claim a certain element of life as off-limits to private or government intervention, and that personal information – when it is used – should be kept confidential. The important point to note for the present investigations is that, in contrast to the discussion in *Privacy and Freedom*, privacy and due process are being presented as fully independent concepts, and that the right of access to personal data is proposed as a safeguard for due process.[32]

The historical background of the importance of constitutional rights in the US can be found in the birth of that republic. Under the influence of political philosophers like Locke and Montesquieu, the intent was to create a state based on the rule of law in which – contrary to the situation in Europe at the time – laws were made by the people, and power was held accountable and was under their control. Probably the most famous system of checks and balances to keep governmental power under control is the separation of powers. In Montesquieu's version, this model consists of separating the legislative, executive, and judicial branches. Due process is another model to control the arbitrary use of power by government and is one of the most valued concepts in US constitutional law.[33] It is codified in the 5th and 14th amendments of the Constitution and is formulated as follows: "No person shall be deprived of life, liberty, property, without due process of law."[34]

While due process was initially applied only to penal cases, over time it developed into a doctrine that applies to other situations in which the government makes a decision that may negatively affect a citizen.[35] The development of due process into administrative law goes hand in hand with the development of the welfare state and the development of theories of positive freedom.[36]

Due process is not only considered a fundamental principle in the US, but also in Europe and in all liberal democracies around the world. The right to a fair trial, for example, which is protected by Article 6 of the European Convention of Human Rights, is also an expression of the principle of due process.[37] Moreover, due process is intimately connected to the concept of rule of law and has similarities with the German Rechtsstaat principle.[38] In the Netherlands, public decision-making power is regulated through the so called "principles of good administration".[39] While the legal systems of due process in Europe and US differ in terms of the exact principles that they incorporate, they share a general structure, and primary function – the control of state power –, and therefore, analyses of due process made with regard to the US system are also relevant for the European context, and their relevance extends to the European debate on data

24 Westin (n 12) 363.
25 While *Databanks in a Free Society* is arguably a more important foundational text for data protection – and in particular for understanding Westin's views on data protection – than *Privacy and Freedom*, the latter is cited around 15 times more frequently than *Databanks in a Free Society*, which can be partly explained by the fact that the first is still in print and easily available, while the second is much harder to find. (A search on google.scholar.com performed in February 2020 yields 287 citations for *Databanks in a Free Society* and 5491 citations for *Privacy and Freedom*).
26 In the language of Westin and Baker, the right of access applies to "files" or "records". The due process view on access (and data protection law) would suggest qualifying personal "records", or personal "files", as "personal data".
27 Westin and Baker (n 1) 355-378.
28 Westin and Baker (n 1) vii and xvii.
29 Westin and Baker (n 1) 5-6.
30 Westin and Baker (n 1) 14-20.
31 Laurence H Tribe, *American Constitutional Law* (2nd edn, The Foundation Press 1988) chapter 10.
32 See also Paul de Hert & Serge Gutwirth, Privacy, data protection and law enforcement. Opacity of the individual and transparency of power, in *Privacy and the Criminal Law* 61 (E. Claes, A. Duff, & Serge Gutwirth eds., 2006). They argue that there is a similar distinction, in the context of European law, between privacy law, which protects the opacity of the individual, and data protection, which mostly channels power through transparency tools.
33 Westin and Baker(n 1) 15 (referring to Justice Felix Frankfurter in *Green v. McElroy, 360, U.S. 474 (1959))*.
34 Due process rights against the federal government are granted in the 5th amendment, while the 14th amendment guarantees due process with regard to states.
35 Tribe (n 31) chapter 10, paragraph 1.
36 Tribe (n 31) chapter 10, paragraph 1.
37 Katja De Vries, 'Privacy, Due Process and the Computational Turn -- A Parable and a First Analysis' in Mireille Hildebrandt and Katja De Vries (eds), *Privacy, Due Process and the Computational Turn* (Routledge 2013).
38 E.g. TRS Allan, 'Freedom, Equality, Legality' in James R Silkenat, James E Hickey Jr. and Peter D Barenboim (eds), *The Legal Doctrines of the Rule of Law and the Legal State (Rechtsstaat)* (Springer International Publishing 2014) https://doi.org/10.1007/978-3-319-05585-5_11. See generally James Silkenat R, James Hickey Jr. E and Peter D Barenboim (eds), *The Legal Doctrines of the Rule of Law and the Legal State (Rechtsstaat) (Springer 2014)* https://doi.org/10.1007/978-3-319-05585-5 noting that obviously, while there are similarities between concepts such as due process, rule of law, Rechtsstaat and principles of good governance, there are also differences.
39 Peter Hendrik Blok, *Het Recht Op Privacy: Een Onderzoek Naar de Betekenis van Het Begrip 'privacy' in Het Nederlandse En Amerikaanse Recht* (Boom Juridische Uitgevers 2002) 118.

protection.[40]

As a set of rules which aims to guarantee the just application of general laws in individual cases, due process is one of the constitutional rights by which power can be held accountable. Moreover, according to Tribe, a prominent scholar of American constitutional law, there is both an instrumental as well as an intrinsic justification for due process. From an instrumental point of view, these rights are indispensable for the exposure of error and bias in adjudication, and they offer the best chance for a procedure to arrive at the truth.[41] From a substantive point of view, these procedures protect human dignity by allowing people to be part of the decision-making process.

The concrete content of the procedural guarantees of due process is composed of numerous elements.[42] Two elements in particular are relevant in the context of processing of personal data: (1) the right to know in advance the evidence of a criminal or administrative case and (2) the right to contest this evidence.

The central policy proposed by Westin and Baker is that due process should be applied to all cases in which judgments are made about people on the basis of their personal records. Giving the citizen a right to access their record is a way to allow them to know and assess how a judgment about them has been reached. Moreover, giving them the right to challenge the record allows them to contest a decision if it has been made on the basis of false or irrelevant facts. These rights should apply to any systematic use of personal records for the same reason as they apply in criminal cases. For example, these rights should apply to files or reports from caseworkers in the case of welfare proceedings. Similarly, they should apply to files in all contexts such as education, clinical psychology, probation, loan decisions. In all these contexts it is important – both for instrumental as well as substantive reasons – that individuals are put in the position to assess and contest the facts and opinions that play a role in the making of decisions about them.

Perhaps surprisingly, Westin and Baker's argument in favor of the introduction of the right of access to personal data is not based on the increased risks posed to a free and democratic society by new forms of digital data processing. They find, based on their empirical work, that the computerization of data processing was in fact not having a negative impact on the rights of citizens.[43] In most cases, rights such as the right to know that a file exists, or the right to access a file about one self, had remained unaltered.[44] Many file-systems existed in the pre-computer era that did not afford these rights, and when the systems containing these files were automated, the same rights were still not granted. At the same time, in fields where people did already have rights to access files and challenge the accuracy, completeness and propriety of the information, these rights were retained when the files got digitized. This was the case, for example, with the rights of social security account holders with regards to their earning records and for veterans with regards to their service records. These examples also remind us that the right of access to files predates the introduction of the right of access within data protection regulation.

The fundamental reason why Westin and Baker argue for the introduction of new rights is the changed public perception and demands for fairness with respect to the exercise of power that was prevalent at the time they were writing, which they believed to be justified. In the 1960's, various movements, including the Civil Rights movement, were seeking a re-balancing of power in society, and demanded a strengthening of civil liberties. They were fighting against many injustices and demanded social, political and legal systems to live up to their professed values of merit selection, equal opportunity and respect for the individual. The movement was critical of "credential-based gate-keeping", disapproving of the extensive data collection and criteria that were used to make decisions about individuals, for example in getting access to housing, jobs and schools. These practices resulted in discrimination, favoring whites over blacks, rich over poor, straight over gay, and in general in the repression of forms of dis-conformity.

Another practice that was heavily criticized was the widespread practice of compiling lists of people showing "deviant" behavior (such as participating in anti-war or anti-discrimination demonstrations), which were used to suppress dissenting opinions. The social unrest of that time came from many different sides, including long-discriminated groups, such as people of color, new sociopolitical groups fighting for women's rights, sexual liberation etc., but also conservative defenders of constitutional principles, and revolutionary groups. Moreover, while the demand for change was led by a variety of activist groups, Westin and Baker show, on the basis of survey data, that the concern for civil liberties issues in relation to privacy and record keeping were held by large segments of the population.[45] In short, against the background of the demands of the civil rights movements, Westin and Baker argue that citizens should finally get, in practice, those rights that so far had only been acknowledged in theory.[46]

While Westin and Baker's argument for the citizens' right of access to their record is primarily based on due process principles, which in first instance protects *individual* citizens' interests, they also see the potential for the right of access to function as a means to mend injustices on the *societal* level. For example, they argue the right to access files enables people to assess whether discriminatory practices are still used. From this perspective, allowing people to assess and criticize how decisions are being made, and safeguarding people's right to (peacefully) dissent are essential to safeguard the functioning of a democratic society.[47]

Westin and Baker argue that in order to transform the ideals underpinning a free and democratic society into enforceable rights, the

40 In their concrete historical development the doctrines of due process did diverge in US and EU law. However, while Westin and Baker refer to some extent to the concrete implementation of the doctrine of due process in US law (and thus to their historically particular specification), their arguments are exclusively based on the core of the concept as it was already developed in the Enlightenment period, and is therefore also applicable to Europe, where due process is also still a fundament of the legal system. *See also Carol Harlow, 'Global Administrative Law: The Quest for Principles and Values' (2006) 17 European Journal of International Law 187, 191* https://doi.org/10.1093/ejil/chi158.

41 Tribe (n 31) Chapter 10, paragraph 7.

42 Blok (n 39) 248. It should be noted that when Westin and Baker talk about due process they refer to what in US legal doctrine is known as procedural due process. It is important to stress this because in the US, certain areas of privacy have found constitutional protection through the application of substantive due process rights. One of the elements of substantive due process is that some interest are protected to such extent that the government is not allowed to interfere with them at all. It is on this basis that the US Supreme Court has ruled to protect diverse "zones of privacy" (Roe v. Wade, opinion of the Court delivered by Justice Blackmun) such as the right to abortion, and the freedom to choose a wedding partner. However, this differs in crucial aspects from the right of access as due process, where it is fundamentally procedural due process that is at play. See Blok (n 39) 178-189.

43 Westin and Baker (n 1) 269.

44 Westin and Baker (n 1) 258.

45 Westin and Baker (n 1) 345.

46 Westin and Baker (n 1) 341-347.

47 Westin and Baker (n 1) 348.

right of access should be extended in three directions. First, "The general principle that should guide the inspection aspect of access legislation is that *any* record about an individual which is consulted by government officials in the determination of the individual's rights, opportunities and benefits under a government program should be open to inspection".[48] No longer should the right be dependent on the particular regulations governing the individual agencies. Second, the right of access should be applied to data not only at the moment when it is used in decision making, but also when data is merely held.[49] Third, the citizens' right to see their record should ideally apply not only to the relationship between government and citizen, but also to the relationship between people and private organizations.[50]

These extensions take the right of access beyond the realm of government decisions that have effect on the life of citizens to which due process is originally applicable. However, as we have seen, the essential function of due process is to act as a mechanism of control on the use of power. Westin and Baker argue that decisions of private entities such as banks and insurance companies have an enormous effect on people's lives and, therefore, a regime similar to the one applicable to government decisions should apply to them. In this regard, they point out that the Fair Credit Reporting Act, which deals with the private sector, paved the way to access laws in the public sector. Expanding the applicability in this direction is important because private decision making is also relevant in the context of the struggle for civil liberties, such as in the fight against discrimination.

One case taken from the work of Westin and Baker illustrates well the concept of right of access to personal data as due process and is paradigmatic for their perspective.[51] The case concerns a woman, Mrs. Tarver, who was receiving welfare support from a state-run aid program. In that context, a civil servant dealing with her case produced a report which included allegations that she abused her child. This report was subsequently handed over to a juvenile court, which had to decide if Mrs. Tarver would lose custody over her children. Mrs. Tarver was ultimately acquitted. Nonetheless she demanded to get access to the file to be able to contest the derogatory information it contained, but the department that held the file denied her access. Therefore, she went to court, with the support of the American Civil Liberties Union, arguing that the file might still be used by other case-workers or other departments. Yet, her request was again rejected. Based on their due process view of the right of access, Westin and Baker argue that Mrs. Tarver should have been granted access to her file in such case, and the ability to contest and correct the information contained in it.[52]

To conclude, Westin and Baker argue for the introduction of a general right to access personal files. The primary aim of this right is to bring the citizen in a position that enables them to judge the veracity and relevance of the image painted of them in a file, and to allow them to contest unfair decisions if necessary. In this perspective, this right is an extension of the doctrine of (procedural) due process – the right to see and contest the evidence brought in criminal cases – towards all situations in which decisions are (or can be) made based on the processing of personal data.

48 Westin and Baker (n 1) 364.
49 Westin and Baker (n 1) 356-357.
50 Westin and Baker (n 1) 371.
51 Westin and Baker (n 1) 357-360.
52 Considering that Westin and Baker refer to Franz Kafka's *The Trial* may also help us understand what harm they want to prevent when arguing for the right of access from a due process point of view.

3. Informational Self-Determination: The Right to Freely Develop Your Own Personality

The right of access, and the other data subject rights, are associated by some scholars with "informational self-determination".[53] Informational self-determination was first introduced into data protection case law as a constitutionally protected right by the German Federal Constitutional Court (Bundesverfassungsgericht) in the 1983 Census decision (Volkszählungsurteil).[54] However, the right of access played only a minor role in that decision, and the court did not directly relate the right of access to informational self-determination. By analyzing the theoretical foundations of informational self-determination, I aim to create a clearer understanding of it, and to show that while the right of access is not at odds with this theory of data protection, it is not one of its central principles.

The background of the Census case is a 1982 census law enacted unanimously by the German parliament (Bundestag), allowing the state to collect detailed demographic information about its citizens through a questionnaire containing over 160 questions.[55] Citizens concerned about their privacy, and other possible risks connected to the increasing role of computers in public administration (including many German data protection scholars such as Podlech, Steinmüller and Brunnstein), challenged the constitutionality of the law in court.[56] The court ruled that the census, including the mandatory nature of participation, was in principle constitutional; nonetheless, it struck down the law for two main reasons. First, the data collected was not only going to be used for statistical purposes, which was the main goal of the census, but also for other tasks of public administration and in branches of government (aside from the federal government). Second, the court held that some of the procedural precautions were lacking detail and needed to be strengthened.[57]

The court ruled that the census, in its proposed form, violated the right to informational self-determination. This new right was derived by this court from the general right to personality, which itself had been developed in previous case law, and was grounded on the right to protection of human dignity (article 1(1) Constitution), and the right to protection of personal liberty (article 2(1) Constitution).[58] The purpose of this right to personality is to guarantee each individual the possibility to freely develop their own personality.[59] The court defined

53 See (n 7).
54 BverfGE 65. See also Gerrit Hornung and Christoph Schnabel, 'Data Protection in Germany I: The Population Census Decision and the Right to Informational Self-Determination' (2009) 25 *Computer Law & Security Review* 84 providing a general description of the case in English.
55 Hornung and Schnabel (n 54) 85.
56 Adalbert Podlech, 'Die Begrenzung Staatlicher Informationsverarbeitung Durch Die Verfassung Angesichts Der Möglichkeit Unbegrenzter Informationsverarbeitung Mittels Der Technik' (1984) 1984 *Leviathan* 85, 91; Jörg Pohle, *Datenschutz und Technikgestaltung: Geschichte und Theorie des Datenschutzes aus informatischer Sicht und Folgerungen für die Technikgestaltung* (Humboldt-Universität zu Berlin 2018) 144 https://www.hiig.de/publication/datenschutz-und-technikgestaltung-geschichte-und-theorie-des-datenschutzes-aus-informatischer-sicht-und-folgerungen-fuer-die-technikgestaltung/.
57 See *Census Decision C.III.2* (a) Citizens had to be proactively informed about their rights such, for example regarding the fact that it was not mandatory to answer to all questions; (b) It should be guaranteed that identifying information would be deleted at the earliest possible moment; (c) There should be strict rules to avoid conflict of interest of those executing the survey; (d) The legislature had to make sure that the actual questions that would end up in the questionnaire are in line with the law.
58 See *Census Decision* C.II "Prüfungsmaßstab ist in erster Linie das durch Art. 2 Abs. 1 in Verbindung mit Art. 1 Abs. 1 GG geschützte allgemeine Persönlichkeitsrecht."
59 Hornung and Schnabel (n 54) 86.

this right to informational self-determination, as a derivative of the right to personality in the context of processing of personal information by the state, as: "the authority of the individual to decide himself, on the basis of the idea of self-determination, when and within what limits information about his private life should be communicated to others."[60] In other words, the court seems to say that processing of personal data should be based on the individual's consent. The reader will note that given this definition, informational self-determination appears to be the same as Westin's notion of "privacy as control".[61] However – and this is essential – the court also ruled that citizens have to accept restrictions to their right to informational self-determination if there is an overriding general interest.[62] Such an overriding general interest was found to be present in the case of a national census and, as a result, the court held that the mandatory character of the census in itself did not infringe unlawfully on the right to informational self-determination.[63]

The most important aspect of the case is that the court discusses the data protection principles which need to be in place in order to protect the right of informational self-determination when processing is not based on consent. The court stresses the principles of transparency and purpose limitation, which serve to ensure that people are aware of the information which is being processed about them. The court supports the need for these principles on the idea that people can only develop freely when they know what other people know about them.[64] In other words, the court asserts that when people are in the condition of not knowing which data is held about them, this constitutes a restriction on their freedom of action. This creates a need for the people to be protected against the unrestricted collection, storage, use and transfer of information relating to them.[65]

It is hard to fully grasp the concept of informational self-determination only on the basis of the deliberations of the court. While the court did not refer to any particular underlying theories behind its decision, its interpretation of the German Constitution did not appear out of thin air. In fact, informational self-determination was being discussed actively in the legal literature in Germany at the time and it seems unquestionable that the court was influenced by this debate. According to German data protection scholars Podlech,[66] and more recently Pohle,[67] the court indeed took the concept and many of its deliberations directly from the academic literature.

Much of the German data protection literature from that time shares one characteristic, namely the fact that legal arguments are developed and grounded in sociological analyses of humans in society, and in particular on sociological systems theory.[68] The work of Luhmann, a sociologist who was one of the developers of that theory, was particularly influential on legal scholars at that time. For example, it influenced Podlech, a prominent data protection scholar, and one of the claimants in the Census case, as well as the PhD thesis of Mallmann, which contains the first clear formulation of the need for a right to informational self-determination.[69] Moreover, a report written by Steinmüller and others (1972) for the ministry of the interior titled *The foundation of data protection* [Grundfragen des Datenschutze – Gutachten im Auftrag des Bundesministeriums des Innern] also relied on sociological systems theory, and on Luhmann's theories more broadly.[70]

Two pillars of sociological systems theory constitute the epistemological precondition of the development of informational self-determination. First – the theory of functional differentiation, according to which modern society should be understood as a system constructed out of a collection of different subsystems (e.g. economic, religious, cultural), each with their own rules, norms, and interactions. The theory further claims that society's ability to progress is dependent on the development of this stratification. Second – sociological role theory, which explains how human beings have to play different roles in these different subsystems. People construct a personality within the confines allowed by the combination of various roles they have in various societal sub-systems. In the words of Luhmann, "every human being is expected to be able to relate his actions to several social systems and to unite their unbalanced demands in a personal synthesis of behavior."[71] It follows from these theories that the success of society as a whole is dependent on the ability of individuals to construct a consistent personality.

In *Constitutional Rights as an Institution*, Luhmann applies his sociological analysis of society to explain the function of the fundamental rights in making possible the functionally differentiated society. He explains that dignity and freedom "describe the basic conditions for the success of a person's self-portrayal as an individual personality."[72] On the one hand, freedom means that there must be aspects of action that do not appear to be directly caused by *external* factors, and therefore can be attributed to the person.[73] Luhmann sees dignity, on

60 See *Census Decision* C.II

61 And indeed the theory of informational self-determination was heavily influenced by Westin's *Privacy and Freedom. See for example Mallmann (n 20) 50-53. See for a more detailed account: Pohle (n 56) 34 and p6.*

62 See *Census Decision* C.II.*1.b)*.

63 See *Census Decision C.III.1.*

64 See *Census Decision* C.II.*1.a)* : "Mit dem Recht auf informationelle Selbstbestimmung wären eine Gesellschaftsordnung und eine diese ermögliche Rechtsordnung nicht vereinbar, in der Bürger nicht mehr wissen können, wer was wann und bei welcher Gelegenheit über sie weiß"

65 See *Census Decision* C.II.*1.a)*: "Hieraus folgt: Freie Entfaltung der Persönlichkeit setzt unter den modernen Bedingungen der Datenverarbeitung den Schutz des Einzelnen gegen unbegrenzte Erhebung, Speicherung, Verwendung und Weitergabe seiner persönlichen Daten voraus."

66 Podlech (n 56) 91 note 4; Henry Krasemann and Martin Rost, Interview with Adalbert Podlech, 'Interviews Zur Geschichte Und Theorie Des Datenschutzes in Deutschland: Podlech' (2008) from 20:32 http://www.maroki.de/pub/video/podlech/interview_podlech_pub_v3_transkription_v1.pdf

67 Pohle (n 56) section 2.4.2.

68 Hornung and Schnabel (n 55) 85; See generally Jörg Pohle, 'Social Networks, Functional Differentiation of Society, and Data Protection' [2012] arXiv:1206.3027 http://arxiv.org/abs/1206.3027. Pohle also argues that data protection is currently generally understood from an individualist perspective, while the purposes of data protection regulation would be better served by applying a structuralist approach.
Data protection theory in Germany was also heavily influenced by the development of cybernetics which studies the dynamics (and stability) of systems as they are regulated through relationships of processing and communication of information. See for example Steinmüller and others (n 71) section 2.2.3, who write that a constitutional foundation for data protection cannot be derived from an understanding of the constitution within itself, but instead should be based on cybernetics and sociology.

69 For instance Mallmann (n 20) chapter 3; Krasemann and Rost (n 66) from 30:22.

70 Wilhelm Steinmüller and others, *Grundfragen Des Datenschutzes Gutachten Im Auftrag Des Bundesministeriums Des Innern* (1972) https://dipbt.bundestag.de/doc/btd/06/038/0603826.pdf.

71 Niklas Luhmann, *Grundrechte als Institution. Ein Beitrag zur politischen Soziologie* (Ducker & Humblot 1965) 53.

72 Luhmann (n 71) 61. With this view, Luhmann criticizes the dogmatic idea of freedom and dignity in the German constitutional tradition, which according to him are tautological. If man is free and has dignity intrinsically from the fact of being man, then they would be in no need of constitutional protection. Instead, these values only have meaning when they are understood from a psychological and sociological perspective.

73 Luhmann (n 72) 66.

the other hand, as the *internal* ability of the person to construct a consistent self-representation. According to him, freedom and dignity are enshrined in the Constitution to protect the conditions that people need for successful self-portrayal.

Informational self-determination should be understood against the background of these conceptualizations of human dignity and freedom, as developed in the context of sociological systems theory. These theoretical foundations explain why individual freedom and dignity, which are required to construct a personality, depend on the capacity of individuals to know the information that other actors in society, including the state, have about them. Against this background, we should also read the Census case and, in particular, the following crucial lines of the court's judgment:

> "Anyone who is not able to oversee with sufficient certainty what information concerning him is known in certain areas of his social environment, and who is not able to assess the knowledge of possible communication partners to a certain extent, can be significantly inhibited in his freedom to plan or decide on the basis of his own self-determination. A social order in which individuals can no longer ascertain who knows what about them and when – and a legal order that makes this possible – would not be compatible with the right to informational self-determination."[74]

With respect to the relation of the right of access to informational self-determination it should be noted that this right was not central in the work of the theorists who developed informational self-determination, and it also played only a minor role in the decision of the court in the Census case. The court saw the right of access as a safeguard for "effective legal protection", not as a safeguard for the protection of informational self-determination.

The right to effective legal protection is granted by Article 19(4) of the German Constitution which states that "where rights are violated by public authority the person affected shall have recourse to law". The claimants argued that the census infringed upon this right because citizens would not be able to know which part of the government would get the information and for which purposes it would be used, and would therefore also not have judicial recourse against these further uses of their data.[75] The court ruled that since the statistical offices were bound to record every transition of data and the citizens had the right to access these records (of the data and of the transmission), the right of the citizens to judicial recourse was sufficiently guaranteed.[76] Interestingly, by understanding the right of access as a means to guarantee the right to effective legal protection, the analysis of the German court could be interpreted as a "due process" understanding of access.[77]

In conclusion, following Schwartz, informational self-determination ought not to be seen as a right of control over personal data, or simply as the German version of "privacy as control".[78] "Privacy as control" is primarily focused on keeping information private and allowing sharing only on the basis of consent, whereas informational self-determination protects people's right to freely develop their personality, by keeping data flows limited, transparent, and geared towards what is necessary for a free and democratic society. The right of access to personal data is not central to informational self-determination, and is instead, also in the German court, understood, in line with the due process view, as a safeguard to effective legal protection.

4. Rodotà: Access as Power Reversal

The final (and crucial) theoretical root of the right of access can be found in the work of Italian scholar Stefano Rodotà. In his 1973 book called *Computers and Social Control* [Italian original: Elaboratori Elettronici E Controllo Sociale], Rodotà explores the kind of legal-institutional framework that would be needed in order to regulate the use of computers and the processing of personal data.[79] I discuss this work here because it offers an understanding of data protection regulation in which the right of access plays a central role. Its central proposition is the collective use of the right of access to personal data as a means to bring structures of power in society under social control. While the book has been one of the foundational texts of the Italian data protection literature,[80] it has received rather limited attention in international scholarship (which now, as well as then, is dominated by the English language).[81] Moreover, while *Computers and Social Control* is informed by the main texts on data protection of the time, including those of Westin, Simitis and Steinmüller,[82] it presents a distinctive angle to data protection, based on a different political-philosophical grounding.

Computers and Social Control is currently not widely known. However, this work, or at least the spirit of the text, arguably had a quite substantial influence on the development of European data protection legislation. Rodotà was a key member of the European data protection policy community, and had a strong influence on data protection theory and practice.[83] He was the first chairman of the Italian Data Protection Authority ("Garante") and was later appointed as chairman of the Article 29 working party. In that capacity, he was a member of the committee that drafted the Charter of Fundamental Rights of the EU,[84] and in this role he proposed by amendment to add

74 *Census decision* C.II.1.a): " Wer nicht mit hinreichender Sicherheit überschauen kann, welche ihn betreffende Informationen in bestimmten Bereichen seiner sozialen Umwelt bekannt sind, und wer das Wissen möglicher Kommunikationspartner nicht einigermaßen abzuschätzen vermag, kann in seiner Freiheit wesentlich gehemmt werden, aus eigener Selbstbestimmung zu planen oder zu entscheiden. Eine Gesellschaftsordnung und die sie ermöglichende Rechtsordnung, in der jemand nicht mehr weiß, wer, wann, was und bei welcher Gelegenheit über ihn weiß, ist mit unserer Verfassung nicht vereinbar."

75 See *Census Decision* A.II.

76 See *Census Decision* C.V.

77 See BJ Goold and others, *Public Protection, Proportionality, and the Search for Balance* (Ministry of Justice 2007). This work discusses the right to effective legal protection in the German Constitution and makes the connection with fair trial rights such as those defined in article 6 of the ECHR.

78 Paul Schwartz, 'The Computer in German and American Constitutional Law: Towards an American Right of Informational Self-Determination' (1989) 37 *The American Journal of Comparative Law* 675, 690 https://lawcat.berkeley.edu/record/1113532/files/fulltext.pdf.

79 Rodotà (n 2).

80 Emilio Tosi, 'High Tech Law in Italy' in Emilio Tosi, *High Tech Law: The Digital Legal Frame in Italy* (Giuffre Editore 2015) 5 http://www.dimt.it/wp-content/uploads/2015/11/Estratto-HTL-Cap.1.pdf.

81 At the time of writing *Computers and Social Control* has 60 citations according to Google Scholar, versus 6594 for Westin's *Privacy and Freedom*. Rodotà refers to in the introduction to the book to a German bibliography of the time which cites 392 texts in English, 15 in German and only 4 in other languages, showing this is a long existing situation (Rodotà (n 2) 7).

82 Rodotà refers to Westin's *Privacy and Freedom* as well as *Databanks in a Free Society* throughout the book.

83 See Bennett (n 3) 128; Lee A Bygrave, 'International Agreements to Protect Personal Data' in James B Rule and Graham Greenleaf (eds), *Global Privacy Protection* (Edward Elgar 2008) 18; Gloria González Fuster, *The Emergence of Personal Data Protection as a Fundamental Right of the EU* (Springer Science & Business 2014) DOI 10.1007/978-3-319-05023-2_3. All highlighting his role and influence on several important committees and expert groups.

84 Article 29 Working Party, 'Fifth Annual Report: On the Situation Regarding the Protection of Individuals with Regard to the Processing of

the right of access to personal data to the Charter, thereby having a crucial role in getting the right recognized as a fundamental right in Europe.[85]

In *Computers and Social Control*, Rodotà discusses the question of whether the existing legal framework for the protection of privacy and confidentiality is able to tackle the most pressing societal problem caused by the processing of personal data by automated means, which he clearly identifies – already at the time – as the accumulation of (economic and political) power vested in the public and private organizations which collect and process personal data. According to his analysis, the main negative effect of the increased use of personal data is the shift of power away from people.[86] On the one hand, organizations have an increasing ability to evaluate and control people through the collection and combination of many sources of personal data. On the other hand, people are less capable of exercising control over the organizations which have the power to control them.

Rodotà provides various practical examples of this control over individuals through the use of personal information: credit rating agencies in the US, as well as in Italy, creating profiles on individuals; the car manufacturer FIAT illegitimately creating personal files on employees, journalists, other industrialists, etc.; intelligence agencies collecting information about politically deviant behavior.[87] One of the crucial dangers of this situation is the psychological deterrent effect that the mere existence of these systems exerts on the behavior of people. He explains that, in the context of credit agencies, for example, an individual may be induced to continue the payment of installments on a faulty product even if it would be legitimate to refuse such payment, because the non-payment may be recorded in the system without the reason behind it, and therefore lead to the refusal of credit in the future. Profiling by intelligence agencies causes people to censor their own political speech and stop legitimate political activity. In the words of Rodotà:

> "The inability to know the places where records can be collected, the possibility of errors or inaccuracies in the data used, as well as the relationships that can be established between the most diverse information and the conclusions that can be drawn from it, all these elements contribute to increasing the fear of the individual towards the new power, all the more intrusive as it is more tied, nowadays, to the acts of daily life."[88]

Rodotà notes that problems around the use of personal data, including the institutionalized power imbalance it creates, existed before the advent of the computer, but that these will be exacerbated by the digital transformation. Contrary to Westin and Baker's conclusion that the introduction of computers will not pose new problems to civil liberties, Rodotà concludes that "the computer does not intervene to corrupt a healthy environment, but to increase and multiply the existing possibilities of abuse."[89] This is in particular the case where an increased centralization of power go hand in hand with the loss of the ability to control those systems of power. Following Klaus Lenk – a German scholar of social informatics – Rodotà expects this to happen for example in the political domain.[90] He quotes Lenk stating that "there may be "vertical" shifts from local government to central government, or, for federal political systems, from State (Land) to federal government. Power might be also shifted horizontally, from the legislative to the executive, form parliament to the government, only the latter having access to large integrated data bases and being abler to make full use of them".[91] Furthermore, he argues that highly expert knowledge is needed to use computers and extract meaningful knowledge from databases, which in turn leads to more technocratic forms of power, and a loss of control for the majority of the people.

The way that Rodotà analyses the problem of data protection derives from a more general critical view on the unequal distribution of power in society, and in particular on the role of the law in maintaining that distribution. Throughout *Computers and Social Control,* Rodotà follows a tradition of fundamental critique of the "bourgeois" legal system.[92] According to this analysis, the legal system generally serves the interests of those that are already in a powerful position, thereby fortifying the prevailing inequalities in society. A key problem addressed by Rodotà, which derives from this tradition, is that the law implicitly presupposes an abstract equality of power between the parties involved, while in reality this equality does not exist. Reality offers plenty of examples in which this equality is clearly fictional: the relation between employer and employee, the citizen and the state, the doctor and the patient, the consumer and the producer, the holder of an electronic database and the person whose data is held.

In Rodotà's view, the then current system of protection of personal information, which focuses on notions of "privacy" and "confidentiality", is not fit to deal with these questions of balance of power, because these concepts originate in private law and, in particular, property law, to which the critique of bourgeois law is primarily directed. In practice, the realm of privacy is erected predominantly to allow the rich and powerful to retain some sphere of autonomy, thus

Personal Data and Privacy in the European Union and in Third Countries Covering the Year 2000, Part II' (W P54, 6 March 2002) 23.

85 See Convention, 'CONVENT 35 Amendments Submitted by the Members of the Convention Regarding Civil and Political Rights and Citizens' Rights' 447-468 https://register.consilium.europa.eu/doc/srv?l=EN&f=ST%204332%202000%20INIT. Rodotà was not alone in proposing this. The group included Jean-Luc Dehaene, (personal representative of Belgium), Kathalijne Buitenweg (Dutch MEP for the Greens), Andrea Manzella (representative of the national parliament of Italy), Piero Melgorani (representative of the national parliament of Italy), Elena Omella Paciotti (Italian MEP for PES), Stefano Rodotà (representative of the government of Italy), Johannes Voggenhuber (Austrian MEP for the Greens).

86 Rodotà (n 2) section 1.5.

87 Rodotà (n 2) sections 1.2 and 1.3.

88 Rodotà (n 2) 16 (Italian original: "L'inconoscibilità dei luoghi dove una documentazione può essere raccolta, la possibilità di errori o inesattezze dei dati utilizzati, le relazioni istituibili tra le più diverse informazioni e le conclusioni che possono esserne tratte: tutto concorre a far crescere il timore dell'individuo verso il nuovo potere, tanto più invadente quanto piú legato, ormai, agli atti della vita quotidiana.").

89 Rodotà (n 2) 21 (Italian original: "Cosi, l'elaboratore elettronico non interviene a corrompere un ambiente sano, ma ad accrescere e moltiplicare le possibilità di abuso già esistenti.") Rodotà attributes the less radical conclusion by Westin to "ideological ambiguities" in his thinking.

90 Rodotà (n 2) section 1.5.

91 Rodotà (n 2) 38 (Italian original "vi sono spostamenti verticali dal governo locale al governo centrale o, per i sistemi politici federali, dagli stati al governo federale. Il potere può inoltre spostarsi orizzontalmente dal legislativo all'esecutivo, avendo quest'ultimo un accesso privilegiato ai dati trattati con l'elaboratore elettronico" Quoting Klaus Lenk, *Automated Information Management In PublicAdministration; Present Developments and Impacts., vol 4* (OECD Publications 1973) 104 https://files.eric.ed.gov/fulltext/ED088463.pdf.

92 This tradition, according to which laws tend to protect the pre-existing power structures in a society, derives from Karl Marx. See generally e.g. Gary Young, 'Marx on Bourgeois Law' in Rita James Simon and Steven Spitzer, *Research in law and sociology: an annual compilation of research, vol 2* (Jai Press 1978). Similar lines of thought are driving the critical legal studies movement (See e.g. Duncan Kennedy, 'The Structure of Blackstone's Commentaries' (1978) 28 *Buffalo Law Review* 205 https://digitalcommons.law.buffalo.edu/buffalolawreview/vol28/iss2/2/.

helping to protect their interests. This can be recognized, for example, in cases where a right to privacy is invoked by wealthy people when the tax authorities want to collect more data to determine their level of income.[93]

Similarly, according to Rodotà, consent as form of regulation fails because it ignores power relations.[94] He argues that consent is illusory as a basis for lawful processing, because there are many cases in which individuals have no real choice but to accept the processing of personal data. This is caused by the fact that there is almost always a pre-existing inequality of power, and by the fact that this inequality is exacerbated by the opaque and specialized nature of electronic data processing. Therefore, according to Rodotà, the existing approach to regulating the processing of personal data, which ignores disparities of power, does not work and only aggravates power inequalities.[95]

As an alternative to the existing framework of privacy and confidentiality, Rodotà proposes a new institutional framework which, instead of presupposing an equality between individuals and controllers of data, takes the imbalance as a starting point and aims at re-balancing it. In order to achieve this, he proposes a series of regulatory tools. His central policy proposition is the expansion of the right to access personal data, as this will give people the ability to assess and contest how data is being used. Moreover, it will more generally improve the ability of people to hold power accountable. According to Rodotà, "The regulation of the right of access imposes a completely new regulation of secrecy and opens up the possibility of new developments in civil rights, expanding the knowledge available to citizens, and thus their power of control over public and private action."[96] Moreover, he contends that the use of computers to manage databases, that were previously manual, can actually make it easier to give people access to data. He writes: "In this way, the power to control the management of information can be extended, theoretically, to each member of a community: once the possibilities of access are extended and spread, this can not only result in a more immediate control over the management of information but will above all affect the control over power that is based on that information."[97]

In order to expand the possibilities of popular control, Rodotà argues that the right of access should go further than just the right of the person concerned regarding their own data. There should be a right to access anonymous as well as aggregated data, and socially relevant statistical and economic data.[98] These types of data are normally held in powerful centralized institutions, only accessible to certain elites and used to exercise control over society. In his view, expanding access to data would mean: "putting citizens in a position to discuss and challenge a considerable share of public and private decisions, operating in less unequal conditions with respect to the holders of the formal power of decision."[99] In this perspective, there is a strong connection between the purpose of the right of access to personal data and the purpose of general freedom of information rights.

Rodotà emphasizes that the right of access to personal data will not, in itself, solve the problem of informational power asymmetry. On the contrary, access, like consent, could paradoxically undermine the position of individuals, by functioning as a way to legitimize processing of personal data, even when the lawfulness of such processing is questionable. In this regard, he writes there can be the objection that "the right of access ends up appearing just as a means of legitimizing the collection and processing of large quantities of personal information, justified by the argument of the possibility of everyone to know the information collected on themselves."[100] Moreover, there is the problem that individual pieces of information, that can be obtained through access requests, do not provide sufficient knowledge and power to the individual. Rodotà warns that, in these ways, a restricted interpretation of the right of access can lead to a strengthening of these existing power structures.

In order to overcome these pitfalls, Rodotà proposes to embed the right of access in a collective framework, a reoccurring point in many dimensions of his analysis. Three collective solutions, which are all aimed at making the social control over institutions of power more effective, stand out. First, Rodotà argues for the establishment of an institution (similar to the current function of independent supervisory authorities) in charge of checking data controllers and looking beyond the claims originating from specific violation of the rights of individual citizens, i.e. looking at the social dimension.[101] Second, Rodotà writes that individual claims for damages are ineffective means of holding data controllers accountable for multiple reasons.[102] The claim for the individual is often too small to make it worth the effort of a legal action. Moreover, even if an individual starts a claim and wins, the economic effect of the individual's single claim is so minor that it won't act as an effective incentive for the controller to structurally change its behavior.

This problem is made worse by the fact that privacy harms are mostly understood only at the individual level, while the societal

93 Rodotà (n 2) 16-17. Conversely, as Rodotà notes and we still see today, we also see how privacy protection is often less for marginalized people, for example in the extended use of personal information and surveillance in the governance of social security systems.
94 Rodotà (n 2) 45-52.
95 Rodotà says that the focus on the formal equality of power of contracting partners clashes with article 3 of the Italian constitution which holds that the State has the obligation to "remove the obstacles of an economic and social nature which, by effectively limiting the freedom and equality of citizens, impede the full development of the human person and the effective participation of all workers in the political, economic and social organization of the country" (Rodotà (n 2) 47). The finding that there are limits to the protection of privacy based on informed consent is currently still widely discussed. See, for example, Solon Barocas and Helen Nissenbaum, 'On Notice: The Trouble with Notice and Consent', *Proceedings of the Engaging Data Forum: The First International Forum on the Application and Management of Personal Electronic Information* (2009) https://nissenbaum.tech.cornell.edu/papers/Big%20Datas%20End%20Run%20Around%20Procedural%20Protections.pdf; Daniel J Solove, 'Privacy Self-Management and the Consent Dilemma' (2013) 126 *Harvard Law Review* 1880 https://harvardlawreview.org/wp-content/uploads/pdfs/vol126_solove.pdf.
96 Rodotà (n 2) 67 (Italian original: "La disciplina del diritto di accesso, dal canto suo, impone una regolamentazione del tutto nuova del segreto e apre la possibilità di nuovi sviluppi dei diritti civili, ampliando le conoscenze a disposizione dei cittadini, e quindi il loro potere di controllo sull'azione pubblica e privata.").
97 Rodotà (n 2) 79 (Italian original: "In tal modo, il potere di controllare la gestione delle informazioni può essere esteso, teoricamente, fino a ciascun membro di una collettività: una volta ampliate e diffuse le possibilità di accesso, ciò può non soltanto risolversi in un più immediato controllo sulla gestione delle informazioni, ma soprattutto incidere sul potere che su quelle informazioni si fonda.").
98 Rodotà (n 2) 115-118.
99 Rodotà (n 2) 120 (Italian original "Quest'ultimo tipo di accesso realizza già una forma di partecipazione, mettendo i cittadini in condizione di discutere e contestare una notevole quota di decisioni pubbliche e private, operando in condizioni di minor disparità rispetto a quelle dei detentori del potere formale di decisione.").
100 Rodotà (n 2) 101 (Italian original: "... il diritto di accesso finisce con l'apparire proprio come un mezzo per legittimare la raccolta e il trattamento di grandi quantità di informazioni personali, giustificate poi con l'argomento della possibilità di ciascuno di conoscere le informazioni raccolte sul suo conto ... ").
101 Rodotà (n 2) 114-115.
102 Rodotà (n 2) 53-55.

harm – which is often bigger and qualitatively different from the mere addition of the individual harms – is mostly overlooked. As a potential strategy to overcome these issues, Rodotà speaks favorably of the American system of class action and proposes to introduce the possibility to also claim non-material damages. Third, he also mentions the problem that the systems of data processing are extremely complex, and that understanding these processes and their connected dangers is, therefore, incredibly difficult if not impossible for the individual citizen.

In more recent work, Rodotà argued that NGOs should be allowed to take up claims for citizens and stressed their role in asking supervisory authorities to investigate cases.[103] While this last proposal is not yet concretely mentioned in *Computers and Social Control*, it clearly resonates with the other ideas that he presented in the book for overcoming the problem of atomization through collective efforts.

Many of Rodotà's proposals to attain this social control are now part of the GDPR. Article 80 GDPR, for example, gives data subjects the right to be represented by an NGO. The fact that several organizations, such as NOYB – European Center for Digital Rights and Privacy International and many others, are now using the right of access to investigate practices of data controllers and to substantiate complaints to supervisory authorities is an indication that a framework, which in line with Rodotà's view, strengthens collective practices has now become central to the governance of data protection in practice.[104]

While Rodotà's thinking cannot be easily tied to one specific school of taught, it is helpful to consider the political and intellectual context in which he is working. After the defeat of fascism, Italy was rife with socio-political experimentation. Rodotà was part of the Radicali Italiani, a libertarian movement, and wrote for their journal *Argomenti Radicali*, in which writings by Noam Chomsky were published as well.[105] Later Rodotà became an independent member of parliament for the Italian communist party and subsequently for its successor, the Democratic Party of the left. Given this background and the content of his work, we may call Rodotà a libertarian socialist, if we define the core postulate of this political stance – following Chomsky – as believing that systems of authority always have the burden of proof upon themselves to demonstrate that they are justified.[106] However, Rodotà seems to have been more focused on finding positions that allowed him to effectively change the legal system, than to dogmatically hold to any particular political philosophy[107]

To conclude, the primary goal of the right of access to personal data in Rodotà's framework is to contribute to attaining social control over the processing of personal data. Starting from the fact that personal data is often collected in situations of power imbalance and used to exert control over citizens, he argues that the right of access should serve as a counterbalance, in particular to place the centers of power under the control of citizens. In order to achieve this, Rodotà proposes a framework that supports collective action.

5. Analysis

Bringing together the historical roots discussed in the previous sections -- with regards to the right of access and to the broader foundations of, and values safeguarded by, data protection in general -- accentuates their relevance in relation to many questions we are facing today.

The historical perspective suggests that the right of access to personal data, is not primarily an expression of the idea of "privacy as control" nor of "informational self-determination". Instead, there are two strong alternative theories/explanations, in both of which access operates as a tool to reverse informational power asymmetries. The right of access generalizes the doctrine of due process, which helps to expose errors and bias, and thereby contributes to correct and just decisions. Moreover, it allows people individually, but also collectively, to contest and confront systems of decision making. In line with the words of Westin and Rodotà (as quoted in the opening epigraphs at the beginning of this text), access to personal data should be seen as a "basic citizen right" or "civil right", which establishes and enables a new way of regulating power in a free society. The framing of the right in these terms is most relevant as it may significantly impact some of the fundamental ongoing discussions on the role and scope of the right of access.

From recent academic literature, as well as from court cases, it is clear that there is no consensus about the purpose of the right of access to personal data. In two recent cases, *Nowak* and *YS and Others*, the European Court of Justice (ECJ) discussed this question.[108] In both cases, the reason data subjects requested access was to allow them to assess and possibly contest the validity of a judgment made about them, but the purpose of the requests is not taken into account in assessing the validity of the requests.

In *Nowak*, the claimant requested access to his exam transcript as well as the comments by the examiner. Kokott, the AG in this case, argued that access should be granted because the purpose of the right of access was not limited to "verify in particular the accuracy of the data and the lawfulness of the processing" as "even irrespective of rectification, erasure or blocking, data subjects generally have a legitimate interest in finding out what information about them is processed by the controller."[109] The due process view on access would allow for an even stronger conclusion in this regard. In fact, it can be argued that the legitimate interest to access personal data exists *especially* when the data subject wants to access information with the purpose of being able to assess and contest a decision made about him or her.

In *YS and Others*, the ECJ has ruled that the purpose of data pro-

103 Stefano Rodotà, 'Of Machines and Men' in Mireille Hildebrandt and Antoinette Rouvroy (eds), *Law, Human Agency and Autonomic Computing* (Routledge 2011) 192.

104 See e.g. Olivia Tambou, 'Lessons from the First Post-GDPR Fines of the CNIL against Google LLC Reports: France' (2019) 5 European Data Protection Law Review (EDPL) 80. Tambou, provides an analysis of a fine by the CNIL against Google, which was the result of a collective complaint of around 1.000 users filed by La Quadrature du Net and NOYB); 'Our Complaints against Acxiom, Criteo, Equifax, Experian, Oracle, Quantcast, Tapad' (Privacy International, 8 November 2018) http://www.privacyinternational.org/advocacy/2426/our-complaints-against-acxiom-criteo-equifax-experian-oracle-quantcast-tapad.

105 See Argomenti Radicali (1977) 1 1 http://bibliotecaginobianco.it/flip/ARG/0100/#2.

106 See Noam Chomsky, *On Anarchism* (Penguin 2014).

107 For example, in its Report on the Charter on Fundamental Rights related to technological innovation, the European Group of Ethics in Science and New Technologies of which Rodotà was a member, argued for including the right of access to personal data to the Charter, primarily with reference to informational self-determination.

108 Case C-434/16, *Nowak*, ECLI:EU:C:2017:994; Case C-141/12, *YS and Others*, ECLI:EU:C:2014:2081. See generally on these cases, See generally on these cases, Nadezhda Purtova, 'The Law of Everything. Broad Concept of Personal Data and Future of EU Data Protection Law' (2018) 10 *Law, Innovation and Technology* 40. https://www.tandfonline.com/doi/full/10.1080/17579961.2018.1452176.

109 Case C-434/16, *Nowak* (Opinion of Advocate General Kokott), ECLI:EU:C:2017:582 para 39.

tection law is to guarantee the protection of the applicant's right to privacy with regard to the processing of data relating to him or her, and that, in line with this, the purpose of the right of access should not be understood as a right to access administrative documents relating to them.[110] However, from a due process perspective of data protection and the right of access, their purpose is clearly not limited to safeguarding privacy. Moreover, from this point of view, the right of access may well be understood as a right to administrative documents, to the extent that those documents apply to the case of one individual person.

The current scholarly debate on "the right to explanation" can also benefit from the historical perspective on the right of access. In particular, the analysis shows that the right of access has always been fundamentally about the right to understand and contest decisions made through the processing of personal data. From this point of view, Article 15 of the GDPR is a precondition for "the right to explanation" (and contestation). More significantly, the right of access, from its genesis, aimed at giving data subjects transparency and the ability to hold power to account. In this perspective, the historical analysis strongly supports the argument made by legal scholars Selbst and Powles that a right of explanation is at the core of Article 15 GDPR.[111] Indeed, if we extend the analogy of procedural principles of justice, which are at the core of the due process view of access rights, many algorithms in today's society can be seen as laws which we do not know the content of, or perhaps even the existence of, and for that reason alone they are intrinsically unjust.

Related to this, the analysis also shows that data protection has a long history of considering that concentrated power can be more effectively regulated through collective practices. This is most evident in the work of Rodotà, according to whom rights given to and exercised by the individual alone will not be enough to empower citizens in practice and can even wrongfully legitimize processing practices, by concealing the underlying power asymmetries. He points out the emancipatory potential of collective action, for example, when he argues for introducing a new right to facilitate class action lawsuits. More importantly, Rodotà states that the system of rights should aim to protect the citizens not as individual units, but as a collective citizenry, empowered to create democratic (popular) control.

It should be noted that collective aspects are also present in the other two traditions. The landmark Census case in Germany was led by a collective of data protection scholars including Podlech, Steinmüller and Brunnstein. And even though the due process view is mostly about protecting the rights of individuals, Westin and Baker clearly situate their call for a generalization of the right of access in the collective demands of the civil rights movement. Also, for them, the right of access provides a first condition to acquire knowledge about the processing of personal data, which also could be used for collective goals such as addressing structural discrimination in decision making.[112]

Lastly, this analysis should also inform how we study the right of access in practice. It found that a fundamental goal of the right of access to personal data is to give citizens a legal tool to confront and contest power in as far as such power depends on the use of personal data. The insight that there is a fundamental imbalance of power between individuals and the actors who wield control through data-intensive systems is the underlying reason for the creation of this right. It would be naive to expect that these strong actors would lose their relative position of power as a result of the mere creation of this right or would willingly comply to the fullest extent without any pushing back. Following this insight, in order to assess the effectiveness of the right of access, we must look at how it functions in these spheres of contestation.

Multiple studies analyze the effectiveness of the right of access to personal data in practice, and the conclusion drawn by most is that its effectiveness is questionable. Organizations often do not uphold the law,[113] they use "discourses of denial",[114] and data subject rights do not function well in increasingly complex digital realities.[115] These conclusions, however, follow from a limited view of the position of the right of access within the larger framework of data protection. We *should not* merely ask if the right is working from a formal legal point of view (e.g. the right does not work, because organizations do not respond within the legally required term). Rather, we should ask if and how the right is *functioning* from a socially embedded point of view that takes account of its inherent nature of means of contestation, and the actual functioning of the right of access. Does the right allow people to meaningfully contest decisions made on the bases of their personal data? Did the balance of power shift in favor of the holder of the right of access as a result of exercising the right?

After having looked at some of the implications of the historical analysis on the right of access. I will now turn to some analysis of the foundations of data protection more broadly. To clarify the distinctions between the four theories that have been discussed Table 1 below elucidates the central governing principles (or focal points) I associate with these theories. Still, I acknowledge that on the level of actual regulation proposed, they have more in common than they differ.[116]

Table 1 The theories of data protection and their central principles

Theory	Central principles
Privacy as control	consent
Informational self-determination	transparency + purpose limitation
Due process	access + rectification + erasure
Social control / power reversal	access + collective action

The central claim of the theory of "privacy as control" is that people should be able to determine for themselves when, how and to what extent information about them is shared with others. Consent is the

110 Case C-141/12 , *YS and Others*, ECLI:EU:C:2014:2081 para 46.

111 Selbst and Powles (n 13). Selbst and Powles argue that articles 13-15 GDPR give a right to "meaningful information to the logic involved" in automated decision making, and this should be interpreted as a right to explanation, meaning, an explanation understandable to data subjects, about the system as well as about individual decisions made by those systems.

112 Westin and Baker (n 1) 371. ("For example, disguised or hidden criteria as to race, sex, political or cultural beliefs, and other discriminatory standards have been declared improper by recent legal enactments: with due process protections [i.e. in context right of access and challenge], individuals will be better able to see whether such criteria are really being rejected in practice.").

113 Mahieu, Asghari and van Eeten (n 14) 17.

114 Clive Norris and Xavier L'Hoiry, 'Exercising Citizen Rights Under Surveillance Regimes in Europe – Meta-Analysis of a Ten Country Study' in Clive Norris and others (eds), *The Unaccountable State of Surveillance: Exercising Access Rights in Europe* (Springer International Publishing 2017) 434-449 https://doi.org/10.1007/978-3-319-47573-8_14.

115 Vrabec (n 7) chapter 4.

116 For example, the right of access and other data subject rights such as erasure and rectification were explored in all these traditions. See, for instance, Steinmüller and others (n 70) 123-126; Rodotà (n 2) 67; Westin and Baker (n 1) 360.

main principle associated with this vision.[117] Informational self-determination includes and builds upon that claim, but further acknowledges and stresses that very often people are not in a condition of deciding when, and under which conditions, information about themselves is communicated to others. Consequently, in all cases when data is communicated, people should be able to know who has access to their personal information, and for which purposes it is used. In order to enable people to have this knowledge, informational self-determination relies on the principles of transparency and purpose limitation. It follows, therefore, that while recent scholarship generally conflates "privacy as control" and "informational self-determination", these are in fact *different*.
Moreover, the notions of informational self-determination and individual control over personal data are often conflated with the notion of economic control over personal data. For example, former European Commissioner for Justice Viviane Reding put the notion of informed consent, the notion of the right to erasure of data for which consent is revoked, and the right to data portability all under the banner of "putting individuals in control of their data".[118] Similarly, prominent legal scholar Orla Lynskey argues that both portability and the right to be forgotten promote informational self-determination.[119] But it is important to note that the purpose of consent and erasure is to give people the right to "determine for themselves when, how and to what extent information about them is communicated to others", *and* therefore allow people to self-present. On the other hand, the purpose of portability is very *different*, i.e., to allow people, as consumers, to more easily switch between services, and increase economic freedom and efficiency. This function of giving people control over data in order to enable economic freedom and competition is new in data protection and has no precedent in the historical justifications of data protection.[120]

Probably the most important insight that the historical analysis provides is that the staunchest proponents of control rights argued for control over data because they believed this was a necessary tool to shift power dynamics by creating the possibility to assess and contest individual decisions as well as systems of decision making. Such a connection is often overlooked in the ongoing debates about data subjects' control, which often focus exclusively on the individual's control over the flow of data as such, and even tends to steer in the direction of data ownership, as this is seen as the ultimate form of control over data.

The object of control for "privacy as control" is the personal data as such, and for "informational self-determination" it is the development of individual personality. Meanwhile, for the "due process" as well as for the "power reversal" perspectives, the primary objects of control are the organizations which engage in the processing of personal data, as well as the decisions they make and the processes they adopt. This shift of perspective is crucial. In fact, by focusing on a restricted understanding of the right of access – as exclusively relating to personal data – we may be falling into what Selbst and others have called a "framing trap", i.e. remaining stuck in a "data frame", while losing sight of other more relevant "socio-technical" or "informational power asymmetry" frames.[121]

Informational power asymmetry, which is getting more and more attention in recent works of our field,[122] has always been a core theme of data protection. In particular, Westin and Baker, as well as Rodotà, propose frameworks for data protection regulation which fundamentally aim at providing a system for balancing power. While Westin and Baker write about "databases", Rodotà about "electronic data processing", and technology has obviously developed a lot since the time of their writing, their central concerns – i.e. the protection of civil liberties, keeping discrimination at bay, and keeping the centralizing of power in new social-technical systems under democratic control – are among the core questions of data protection today. It would be no exaggeration to say that balancing of power has been the central justification for data protection and is still the very reason for the existence of the GDPR.

6. Conclusion

A key message of this paper is that the different theories of data protection are grounded in different scientific discourses, which conceptualize the relationships between knowledge, law, technology and power in different ways, and therefore offer significantly different justifications for data protection.

Westin and Baker start from the value of due process, a central aspect of constitutional thinking, rooted in the enlightenment. The central idea is that if knowledge has to be produced about the individual, then the individual has to be part of the knowledge production as a form of counter-power. With the increasing use of data, particularly in the context of the welfare state, the due process procedures, understood as a way of regulating power/knowledge relations, are then used beyond the domain of penal law where they originated. Here we see, in short, how an existing form of truth production is translated into a new socio-technological situation.

The German tradition of informational self-determination, in contrast, originates from the then nascent field of sociology. Here we observe how the concepts of liberty and autonomy evolved when man turned the gaze upon himself. With the development of new forms of knowledge (namely sociology), subjectivity is introduced in defining fundamental rights; people should have the right to know and control which data is held about them, because their free self-development is understood to be conditional on knowing this.

Stefano Rodotà, lastly, looks directly at the effects that technolog-

117 In *Privacy and Freedom*, Westin stresses the importance of consent, and several authors such as Barocas and Nissenbaum associate privacy as control mainly with consent. Barocas and Nissenbaum (n 95) 45 ("allowing information subjects to give or withhold consent maps onto the dominant conception of privacy as control over information about oneself"); See Joris Van Hoboken, 'The Privacy Disconnect' in Rikke Frank Jørgensen (ed), *Human Rights in the Age of Platforms* (MIT Press 2019) 265-269 https://www.ivir.nl/publicaties/download/privacy_disconnect.pdf. It should be noted that these authors criticize a simplistic "privacy as control" understanding of the fundamental right to data protection or privacy.

118 Reding (n 6) 124-126.

119 Lynskey (n 10) 591 ("The additional rights granted to individuals by data protection, such as the right to data portability, allow individuals to better determine how their data is processed, by whom and for what purposes. In other words, they promote informational self-determination.").

120 See generally James Meese, Punit Jagasia and James Arvanitakis, 'Citizen or Consumer? Contrasting Australia and Europe's Data Protection Policies' (2019) 8 *Internet Policy Review* https://doi.org/10.14763/2019.2.1409. This article provides an excellent analysis of the values underlying the introduction of a right to data portability.

121 See generally Andrew D Selbst and others, 'Fairness and Abstraction in Sociotechnical Systems', *Proceedings of the Conference on Fairness, Accountability, and Transparency* (ACM 2019) https://doi.acm.org/10.1145/3287560.3287598.

122 See e.g. Lynskey (n 10) 592-597: Damian Clifford, Inge Graef and Peggy Valcke, 'Pre-Formulated Declarations of Data Subject Consent—Citizen-Consumer Empowerment and the Alignment of Data, Consumer and Competition Law Protections' (2019) 20 *German Law Journal* 679, 682 https://doi.org/10.1017/glj.2019.56.

ical change is having on the structures of power in society. In his analysis, the technological development, when left unchecked and only governed by the existing "bourgeois" private law conception of privacy, reinforces existing power imbalances. However, according to him, technology can also lead to new emancipatory practices when supported by new legislation – It is in fact through the collective right of access to data that knowledge, and thereby power, can be redistributed.

The right of access to personal data historically originates neither from the concept of "privacy as individual control", nor from the related but different concept of "informational self-determination". Instead, seen from the due process view developed by Westin and Baker, the right of access allows people to be involved in and question decisions made on the basis of data about them. Such a right derives from a longstanding western constitutional tradition to empower citizens against unjust and incorrect decisions. According to the critical tradition in which Rodotà was situated, people are in a structural power imbalance with regard to state institutions, as well as corporations. In this view, the right of access is a tool to contest the opacity of systems of power and to bring about a higher level of popular control over these systems.

The introduction of elements in the GDPR that flank the right of access to personal data – such as clearer rules for transparency, stronger enforcement capabilities by data protection authorities, and the explicit recognition of the role of civil society actors – brings European data protection regulation much closer to Rodotà's ideal of setting in motion a reversal of power from hegemonic systems back to citizens.

Acknowledgments

I thank Ilaria Buri, Joris van Hoboken, Hadi Asghari, Paul de Hert, Mirna Sodré de Oliveira, Jörg Pohle, Simone Casiraghi, and the anonymous reviewers for their critical remarks and support in writing this paper.

06

Technology and Regulation

Special Issue: Should Data Drive Private Law

Fostering Consumer Protection in the Granular Market:

the Role of Rules on Consent, Misrepresentation and Fraud in Regulating Personalized Practices

Antonio Davola*

Consumer protection, GDPR, consent

a.davola@uva.nl

adavola@luiss.it

Companies increasingly employ data-driven technologies for the allocation and display of offers and advertising based on detailed consumer monitoring. Consumers may fail to recognize the manipulation of their choices if they are unaware of the exploitation of their habits, mental models, and biases. Companies may make use of consumers' cognitive limitations and individual frailties to their disadvantage. Against this backdrop, private law rules could provide meaningful normative guidance in regulating personalized commercial practices.

The article examines the role and characteristics of provisions regulating defective consent and misrepresentation to evaluate whether these rules could incorporate emerging findings on personalized practices and operate as viable instruments for the modernization of consumer protection.

1. Personalized practices in the digital environment

It is commonly understood that, in recent years, online commerce has experienced a profound technological revolution, gradually shifting towards the intensive use of automated data-driven technologies for the allocation and display of offers and advertising for consumers. The ceaseless introduction of tracking and targeting technologies that leverage consumer data in order to personalize the marketing experience has been a defining feature of the impressive growth of online markets.[1] The ability to scrutinize the interests, motivations and needs of consumers through profiling algorithms is at the very core of new modes of creating and supplying products and services in a digital environment.[2]

These innovations provide companies with new ways to gain market advantage. By connecting and cross-examining data obtained from consumers through different sources –e.g. the use of information and communications technologies (ICT), technologies for the internet of things (IoT), and even merely monitoring online activity – companies can intensely scrutinize their (actual and potential) customers and even manage to induce their emotions through affective computing analysis, in order to provide highly personalized offers.[3] This process goes under the general name of 'customerization' and combines both operational and interactional flexibility to tailor not only the product offered, but every aspect of the consumption experience. Online customerization affects both product components (namely their attributes and benefits) and their presentation, choice, and delivery, which in turn impacts the general interaction between the communicator (the seller) and the communicant (i.e. the consumer).[4]

Tailored and targeted commercial techniques constitute a heterogeneous phenomenon and can be based on a vast set of theoretical and methodological underpinnings. Well-known strategies incorporate ex multis semantics and data mining stemming from artificial intelligence,[5] auction theory, and social network and neuroscientific analyses.[6] In addition, they rely on self-tuning algorithms, intent data and immersive multimedia[7] to reach different degrees of personaliza-

1 Alan Schwartz, 'Legal Implications of Imperfect Information in Consumer Markets' (2004) 151(1) *Journal of Institutional and Theoretical Economics* 31, 38.

2 See Irina Domurath, 'Technological Totalitarianism: Data, Consumer Profiling, and the Law' in Lucila de Almeida, Marta Cantero Gamito, Mateja Durovic and Kai Purnhagen (eds.) *The Transformation of Economic Law: Essays in Honour of Hans-W. Micklitz* (Hart 2019), 66: 'Profiling is a term from information science that refers to the construction and application of user profiles through computerised data analysis, increasingly involving the processing of large quantities of aggregated data. During the profiling process, data is analysed and evaluated with the help of algorithms or heuristics, and the constructed profiles are applied as a basis for a decision-making'.

* Antonio Davolo Ph.D. (Sant'Anna School), LL.M. (YLS), is a Postdoctoral Fellow (Luiss), Marie Curie Research Fellow, University of Amsterdam, Fair Personalization Project.

Received 3 Dec 2020, Accepted 5 May 2021, Published 10 Dec 2021.

3 Rafael Calvo, Sidney D'Mello, Jonathan Gratch and Arvid Kappas (eds.), *The Oxford Handbook of Affective Computing*, (OUP 2014); see also Lee Jonathan Steen and Robert Morris Kim, 'Affective Computing: Invasive Technology and Legal Considerations to Protect Consumers' (2010) XI(1) *Issues in Information Systems.*

4 *Ex multis* Soontae An, Hannah Kang and Hyun Seung Jin, 'Self-Regulation for Online Behavioral Advertising (OBA): Analysis of OBA Notices' (2018) 24 *Journal of Promotion Management* 270-291.

5 Bernhard Anrig, Will Browne and Mark Gasson, 'The Role of Algorithms in Profiling' in Mireille Hildebrandt and Serge Gutwirth (eds.) *Profiling the European Citizen – Cross-Disciplinary Perspectives* (Dordrecht 2018).

6 Fabiana Di Porto and Mariateresa Maggiolino, 'Algorithmic Information Disclosure by Regulators and Competition Authorities' (2019) *Global Jurist.*

7 Natali Helberger, 'Profiling and targeting consumers in the Internet of

tion. Such methods have been categorized by scholars under different names, referring inter alia to online behavioural advertising (or OBA),[8] psychological targeting,[9] personalized commercial practices,[10] and micro-targeting.[11]

Amongst legal scholars, growing debate has subsequently arisen on whether and how these techniques should be regulated under the European framework, in reference to different bodies of law – e.g. data protection and consumer law - depending on the specific risk considered.[12] Yet little attention has been devoted to investigating the role that private law can play as a resource in protecting individuals against the threats that highly personalized practices may introduce. In contrast, this article argues that the sector-specific regulations frequently evoked as a means to respond to the personalization of product offers and advertising present shortcomings in term of dealing with the systemic effects of this phenomenon, and that private law rules can constitute an effective resource to enhance consumer protection. In particular, the argument is made that rules on defective consent can provide a valid resource to monitor, scrutinize and correct possible adverse effects arising from personalized techniques.

It should be noted that targeted commercial practices, and customerization more generally, are supposed to introduce significant benefits for all participants in the market ecosystem. From a theoretical perspective, the ability to accurately profile customers improves the market's capacity to match buyers and sellers, therefore lowering both search and transaction costs for products and services.[13] In addition, gathering data from consumers and using consumers as 'informative agents' supports the provision of free online content for the public, in accordance with the paradigm of data as counter-performance.[14] On the whole, it is therefore conventionally acknowledged that a virtuous employment of targeting processes is likely to stimulate economic growth and welfare in the digital sector.[15]

At the same time, the uncontrolled use of consumer data to elaborate predictive and explicit profiles,[16] i.e. used to develop targeted strategies, is likely to lead to manipulations in both quantitative and qualitative terms. On the one hand, this might be the result of a company taking advantage of the asymmetric information that emerges from the elaboration of data gathered for customer classification and profiling – and thereby favouring its resulting traditional market failures; on the other hand, deep knowledge of consumer characteristics might make it possible to influence their choices and exploit their cognitive limits and biases,[17] causing 'behavioural market failures'.[18] In such cases, consumers exposed to tailored commercial offers could end up being unable to recognize the artificial modulation of their set of choices and, possibly, the means available to oppose it, because they are unaware of the way products, offers, and advertisements use their habits, mental models, and heuristics to influence their behaviour.

The result of these and related trends is that, via personalized practices, firms are not only capable of taking advantage of their general understanding of consumers' cognitive limitations but are also able to reveal, and even trigger, the frailties of consumers at an individual level, thus granularizing their business approach depending on the counterpart's characteristics.[19] At the same time, profiles can be used to offer products to specific target groups (or individuals) only, thereby excluding other consumers from access and purchase – or subjecting them to different conditions.[20]

Due to the inner ambiguity of its uses, the growth of profiling as a standard mode of business operation[21] has been viewed with suspicion by scholars and regulators in light of the development of the Digital Services Act package,[22] with some parties calling for the introduction of stringent regulations (that could ultimately favour less intrusive forms of advertising that do not require extensive tracking of user interaction with content)[23] and even promoting a ban on such practices.[24] Currently, though, no explicit option in favor of general

Things – A new challenge for consumer law' in Reiner Schulze and Dirk Staudenmayer (eds.) *Digital Revolution: Challenges for Contract Law in Practice* (Nomos 2016), 135-161.

8 Frederik Zuiderveen Borgesius, 'Online behavioral advertising: a literature review and a research agenda' (2017) 46 *J Advert* 383-376; Sandra Wachter, 'Affinity profiling and discrimination by association in online behavioural advertising' (2020) 35(2) *Berkeley Tech Law J*; Steven C. Bennet, 'Regulating online behavioral advertising' (2010) 44 *J Marshall Rev* 899.

9 Sandra C. Matz et al., 'Psychological targeting as an effective approach to digital mass persuasion' (2017) 114 *Proc Natl Acad Sci* 12714-12719.

10 Przemysław Pałka, Agnieszka Jabłonowska, Hans-W. Micklitz and Giovanni Sartor, 'Before machines consume the consumers. High-Level Takeaways from the ARTSY Project' (2018) *EUI Working Papers*, LAW 2018/12 2.

11 Martin Ebers, 'Beeinflussung und Manipulation von Kunden durch „Behavioral Microtargeting" (2018) *MMR* 7.

12 See *infra* Section 3.

13 Alisa Frik, Amelia Haviland, Alessandro Acquisti, 'The Impact of Ad-Blockers on Product Search and Purchase Behavior: A Lab Experiment' (2020) *USENIX Security Symposium* 163-179.

14 The notion of 'informative agents' was developed by Luciano Floridi, *The Fourth Revolution: How the Infosphere is Reshaping Human Reality* (OUP 2014), 77. See also, for an analysis of data as counter-performance, Sebastian Lohsse, Reiner Schulze and Dirk Staudenmayer (eds.) *Data As Counter-Performance - Contract Law 2.0?* (Hart 2020).

15 Bart Custers, 'Data Dilemmas in the Information Society: Introduction and Overview', in Bart Custers, Toon Calders, Tal Zarsky and Bart Schermer (eds.), *Discrimination and Privacy in the Information Society: Data Mining and Profiling in Large Databases* (Springer 2013), 14.

16 See Article 29 Data Protection Working Party, 'Opinion 2/2010 on online behavioural advertising' (2010) 00909/10/EN WP 171 5 http://ec.europa.eu/justice/data-protection/article-29/documentation/opinion-recommendation/files/2010/wp171_en.pdf.

17 Ryan Calo, 'Digital Market Manipulation' (2013) 82 *George Wash Law Rev* 995.

18 Oren Bar-Gill, *Seduction by Contract: Law, Economics, and Psychology in Consumer Markets* (OUP 2012) 2-4; Cass R. Sunstein, 'The Storrs Lectures: Behavioral Economics and Paternalism (2013) 122 *The Yale Law Journal* 1834.

19 Hans W. Micklitz, 'De- or Re-typification through Big Data Analytics? The Case of Consumer Law' in Christoph Busch and Alberto De Franceschi (eds.) *Algorithmic Regulation and Personalized Law. A Handbook* (Hart 2020); also, Rossella Incadorna and Cristina Poncibò 'The average consumer, the unfair commercial practice directive, and the cognitive revolution' (2007) 30 *J of Cons Policy* 1 21-38.

20 Wachter (n 8), 5.

21 Meike Kamp, Barbara Körffer and Martin Meints, 'Profiling of Customers and Consumers - Customer Loyalty Programmes and Scoring Practices', in Mireille Hildebrandt and Serge Gutwirth (eds.) *Profiling the European Citizen* (n. 8) 201.

22 Proposal for a Regulation of the European Parliament and of the Council on a Single Market For Digital Services (Digital Services Act) and Amending Directive 2000/31/EC COM/2020/825 final (DSA) and Proposal for a Regulation of the European Parliament and of the Council on contestable and fair markets in the digital sector (Digital Markets Act) COM/2020/842 final (DMA), https://eur-lex.europa.eu/legal-content/en/TXT/?uri=COM%3A2020%3A842%3AFIN.

23 See EU Parliament Committee on the Internal Market and Consumer Protection, 'Report with recommendations to the Commission on the Digital Services Act: Improving the functioning of the Single Market' 2020/2018(INL) (2020) https://www.europarl.europa.eu/doceo/document/A-9-2020-0181_EN.html.

24 EU Parliament Committee on Legal Affairs, 'Report with recommendations to the Commission on a Digital Services Act: adapting commercial and civil law rules for commercial entities operating online' 2020/2019(INL) (2020) https://www.europarl.europa.eu/doceo/document/A-9-2020-0177_EN.html: the committee 'invites the Commission to assess options for regulating targeted advertising, including a phase-out leading to a prohibition'.

prohibition is present in the proposal, and targeted advertising is addressed only by means of transparency duties. In particular, online platforms that display advertising are required to make a repository publicly available (through application programming interfaces) that contains information on the aggregate numbers for groups of recipients to whom a personalised advertisement is specifically targeted.[25] The proposal is expected to undergo further modifications and a margin of improvement is definitely present. However, the option of introducing a veto on targeted advertising appears neither feasible nor advisable if we consider – along with the risk of their potential misuse – the abovementioned counterbalancing benefits of these technologies in terms of consumer empowerment and the promotion of prosumerism.[26]

A critical approach to the regulation of targeted practices shall, as a consequence, start from the risks that these techniques pose for consumers. Given this background, the paper investigates the role and characteristics of private law rules regulating consent and misrepresentation as resources to incorporate emerging findings on personalized practices, and evaluates their role as viable instruments for the modernization of consumer protection.

Accordingly, the article first provides an overview of the risks arising from targeted practices, examining them using the common conceptual framework of discrimination (Section 2). By distinguishing different discriminatory harms arising from these techniques, it is possible to highlight the limits of the different regulations that legal scholars have investigated as prospective tools for the phenomenon. Particular attention is devoted to exposing the shortcomings of rules on data protection, competition law, and consumer protection when addressing personalized practices, especially where the problem of reduced consumer self-determination is considered (Section 3).

Following on these considerations, the role of European private law and its interaction with consumer protection is then investigated. This paper argues that provisions on defective consent might constitute a viable regulatory solution, providing a tool to enhance consumer protection and promote substantive social justice in personalized interactions (Section 4). Building on the model rules from the Principles of European Contract Law and in the Draft Common Frame of Reference, the article highlights the view that, conceptually, Member States' rules on defective consent share conceptual ground with the main existing regulatory solutions usually considered when attempting to tackle the risks around tailored commercial practices. In addition, these rules overcome the current limits faced by each of them and therefore can providing a potentially more effective resource for dealing with the phenomenon.

Lastly, the paper offers some considerations regarding further advisable developments in the European framework (Section 5). In particular, a major obstacle is found in the persisting tensions between national and EU principles of contract law. The need for further harmonisation of European principles of contract law is identified as a desirable means to reach a common understanding of social justice in Europe as well as a way to attenuate, integrate and correct adverse and discriminatory effects arising from targeted practices.

25 See DSA Article 30 'Additional online advertising transparency'.

26 See *inter alia* Christian Thorun and Jane Diels, 'Consumer Protection Technologies: An Investigation into the Potentials of New Digital Technologies for Consumer Policy' (2020) 43 *J of Cons Policy* 178; Veronica Marotta, Kaifu Zhang and Alessandro Acquisti, 'The Welfare Impact of Targeted Advertising' (2017) https://ssrn.com/abstract=2951322 or http://dx.doi.org/10.2139/ssrn.2951322.

2. Targeted practices, discrimination and self-determination

Targeted practices raise a plurality of legal challenges, undermining different rights to which individuals are entitled in the digital environment. In the recent literature addressing this topic, risks have often been grouped under the general umbrella notion of 'discrimination.'[27] In this context – and in contrast with its sector-specific meaning in non-discrimination law[28] – the term discrimination is employed according to its descriptive definition, building on its etymologic roots[29] and without implying a structural relationship with protected factors.[30] In spite of using a unitary concept, however, discriminatory effects can be expressed in (at least) three different forms, and these have been unevenly examined in scholarly debate.

A first – and extensively investigated – form of discrimination arising from targeted commercial practices involves the possible exploitation of consumers' cognitive biases[31] and heuristics to exercise undue influence[32] and trigger desired behaviours in the transaction process.[33] In these cases, profiling is implemented to take advantage

27 *Ex multis* Angelisa Plane, Elissa Redmiles, Michelle Mazurek and Michael Carl Tschantz, 'Exploring User Perceptions of Discrimination in Online Targeted Advertising' (2017) *Proceedings of the 26th USENIX Security Symposium*; Wachter (n 8); Nizan Geslevich Packin and Yafit Lev Aretz, 'Social Credit And The Right To Be Unnetworked' (2016) 2 *Columbia Business Law Review*; Solon Barocas and Andrew Selbst 'Big data's disparate impact (2016) 104 *California Law Rev* 671; Pauline Kim, Data-driven discrimination at work (2016) 58 *Wm & Mary L Rev*, 857; Joshua Kroll, Solon Barocas, Edward Feltenm, Joel R Reidenberg, David Robinson and Harlan Yu, 'Accountable algorithms' (2016) 165 *U Pa L Rev* 633; Frank Pasquale and Danielle Citron 'Promoting Innovation While Preventing Discrimination: Policy Goals for the Scored Society' (2014) 89 *Washington Law Review* 1413; John Wihbey, 'The possibilities of digital discrimination: Research on e-commerce, algorithms and big data' (2015) *Journalist Resource* https://journalistsresource.org/studies/society/internet/possibilities-online-racial-discrimination-research-airbnb.

28 Defining cases in which a decision occurs on the sole basis of the parties' protected factors, such as sex, race, ethnic origin, disabilities, religion or belief, age and sexual orientation; see Sandra Wachter, Brent Mittelstadt and Chris Russel, 'Why Fairness Cannot Be Automated: Bridging the Gap Between EU Non-Discrimination Law and AI', *Computer Law & Security Review* 41 (2020) 105567 https://papers.ssrn.com/sol3/papers.cfm?abstract_id=3547922. In this sense, unlawful discrimination can also emerge as the result of personalization processes exploiting protected factors – e.g. gender-based distinctions. See Martin Ebers, 'Regulating AI and Robotics: Ethical and Legal Challenges' in Martin Ebers and Susana Navas (eds.) *Algorithms and the Law* (CUP 2020) 76.

29 The notion comes from the Latin term *discrimen* (distinction) and from the verb *discernere* (distinguish).

30 See Andrew Altman, 'Discrimination' (2020) *Stanford Encyclopedia of Philosophy*.

31 Ariel Ezrachi and Maurice E. Stucke 'The rise of behavioural discrimination' (2016) 37(12) *European Competition Law Review* 485-492; John Hanson and Douglas Kysar, 'Taking Behavioralism Seriously: Some Evidence of Market Manipulation' (1999) 112 *Harvard Law Review*, 1447.

32 *Ex multis* Martha Chamallas, 'The Disappearing Consumer, Cognitive Bias and Tort Law' (2014) 6(1) *Roger Williams University Law Review* 34; Thomas Gilovich, Dale Griffin and Daniel Kahneman, *Heuristics and biases: The psychology of intuitive judgment*, (CUP 2002); Christine Jolls and Cass Sunstein, 'Debiasing Through Law' (2006) 35 *Journal of Legal Studies*; Govind Persad, 'When, and How, Should Cognitive Bias Matter to Law?' (2014) 32 *Minnesota Journal of Law and Inequality* 103.

33 Giovanni Sartor, 'New aspects and challenges in consumer protection. Digital services and artificial intelligence' (2020) EU Policy Department for Economic, Scientific and Quality of Life Policies studies https://www.europarl.europa.eu/RegData/etudes/STUD/2020/648790/IPOL_STU(2020)648790_EN.pdf; see also Agnieszka Jabłonowska, Maciej Kuziemski, Anna Maria Nowak, Hans-Wolfgang Micklitz, Przemyslaw Palka and Giovanni Sartor, 'Consumer law and artificial intelligence: challenges to the EU consumer law and policy stemming from the business' use of artificial intelligence: final report of the ARTSY project (2018) *Working Pa-*

of cognitive limitations characterizing a target group[34] in order to stimulate them to purchase products or services they would otherwise not be willing to acquire (or, at least, that they would shop for under different conditions)[35] or to diversify prices for products and services according to individuals' willingness to pay.[36] Although perfect price discrimination is often thought to be welfare-enhancing by making it possible to achieve efficient outcomes in the distribution of resources, third-degree price discrimination – i.e. charging different segments of the market different prices for the same product, directly linking prices to consumers' willingness and ability to pay – based on exogenous identifying features is also likely to lower consumer welfare by favouring companies' extraction of information rents.[37] In addition, these risks are further exacerbated in concentrated markets such as that of the IoT, with GAFAM[38] operating as oligopolists across industries.[39]

A second established narrative investigates targeted commercial practices as a potential threat to privacy and data protection rules. From this perspective, attention has been devoted to investigating the potential use of sensitive data encompassing protected factors to provide personalized services (both directly and indirectly, or by association),[40] with major consequences in terms of disparate impact.[41] Profiling has been under intense scrutiny by privacy advocates, as well as being normatively addressed by the General Data Protection Regulation (GDPR).[42] Moreover, various proposals have been formulated by scholars in order to regulate this activity[43] and ensure that consumers are able to protect their privacy in the automated processing of their data by digital platforms.[44] With regard to the potential discriminatory effects of profiling, Art. 22 GDPR shall be read in conjunction with the general prohibition regarding special categories of personal data in Art. 9 GDPR, which regulates the processing of personal data items that reveal protected factors and mandates human supervision, provided the exempting conditions set out in the provision do not apply.

Lastly, a third strand of research exists. Namely, personalized commercial practices can be analysed as techniques that affect a consumer's freedom of choice by artificially modulating the sets of products offered on the market; consequently, they can operate as tools for the indirect reduction of consumer autonomy. It has been empirically observed that personalization affects clickthrough rates, and exposure to tailored offers increases user propensity to conduct both active and passive searches on advertiser webpages. Nonetheless, while the impact of these techniques on acquisition rates has been measured by looking at metrics such as purchase probabilities, sales, and online searches,[45] little attention has been devoted to the analysis of the manner in which personalized and behavioural practices undermine self-determination in business-to-consumer transactions.[46]

With personalized practices, consumers exposed to them only see a minor (individually created) subset within the whole assortment of products of the same kind that are present on the market; in addition,

per, EUI LAW, 2018/11 https://cadmus.eui.eu/handle/1814/57484; Christopher Burr and Nello Cristianini, 'Can machines read our mind?' (2019) 29 *Minds and machines* 461-494. It should be noted that critiques of behavioral manipulation are generally value-neutral, meaning that subliminal influence is considered harmful, even when it is meant to achieve legitimate ends: see Cass Sunstein and Lucia Reisch, 'A Bill of Rights for Nudging' (2019) 8(3) *Journal of European Consumer and Market Law* 95.

34 Raffaele Caterina, 'Psicologia della decisione e tutela del consumatore' (2012) 1 *Analisi Giuridica dell'Economia* 2-18; Anne-Lise Sibony and Geneviève Helleringer, 'EU Consumer Protection and Behavioural Sciences: Revolution or Reform?' in Alberto Alemanno and Anne-Lise Sibony (eds.) *Nudge and the Law: A European Perspective* (Hart Publishing 2015), 209-234; Hans-Wolfgang Micklitz, Lucia Reisch and Korlenia Hagen, 'An Introduction to the Special Issue on 'Behavioural Economics, Consumer Policy, and Consumer Law' (2011) 34 *Journal of Consumer Policy* 271.

35 For a general overview, see Sophie Bienenstock, 'Consumer Bias', in Alain Marciano and Giovanni Battista Ramello (eds.) *Encyclopedia of Law and Economics* (Springer 2018).

36 *Ex multis* Inge Graef, 'Algorithms and fairness: what role for competition law in targeting price discrimination towards end consumers?' (2018) 24(3) *Columbia Journal of European Law* 541-559; Ariel Ezrachi and Maurice Stucke, *Virtual Competition. The Promise and Perils of the Algorithm-Driven Economy* (HUP 2016); Mariateresa Maggiolino, 'Personalized prices in European competition law' (2017) *Bocconi Legal Studies Research Paper* https://papers.ssrn.com/sol3/papers.cfm?abstract_id=2984840.17; Antonio Davola, Technological innovation in creditworthiness assessment (2019) 10 Open Review of Management, Banking and Finance. This phenomenon is often given the name 'behavioral exploitation' as well. See Peter Rott, 'A Consumer Perspective on Algorithms', in Lucila de Almeida, Marta Cantero Gamito, Mateja Durovic and Kai Purnhagen (eds.) *The Transformation of Economic Law: Essays in Honour of Hans-W. Micklitz* (Hart 2019), 43-64, 46; Salil Mehra, 'Algorithmic Competition, Collusion, and Price Discrimination' in Woodrow Barfield (ed.), *The Cambridge Handbook of the Law of Algorithms* (CUP 2020), 199-208.

37 *Ex multis* see Gerrit de Geest, *Rents: How Marketing Causes Inequality* (Beccaria 2018) *passim*.

38 GAFAM stands for Google, Apple, Facebook, Amazon, and Microsoft.

39 See Nicolas Petit, *Big tech and the digital economy: The Moligopoly scenario* (OUP 2020).

40 Domurath (n 2), 86; Catalina-Adriana Ivanus, 'Discrimination by Association in European Law' (2013) 2 *Persp Bus LJ* 117.

41 Frederik Zuiderveen Borgesius, 'Personal data processing for behavioural targeting: which legal basis?' (2015) 5 *Int Data Priv Law* 163-176; Chris Hoofnagle, Ashkan Soltani, Nathan Good, Dietrich James Wambach and Mika D Ayenson, 'Behavioral Advertising: The Offer You Cannot Refuse' (2012) 6 *Harvard Law & Policy Review* 273; Maja Brkan, 'Do Algorithms Rule the World? Algorithmic Decision-Making in the Framework of the GDPR and Beyond' (2019) 27(2) *International Journal of Law and Information Technology* 91–121; Barocas and Selbst (n 27); Natalia Criado and Jose M. Such, 'Digital Discrimination', in Karen Yeung and Martin Lodge (eds.), *Algorithmic Regulation* (OUP 2019), 87.

42 Regulation (EU) 2016/679 of the European Parliament and of the Council of 27 April 2016 on the protection of natural persons with regard to the processing of personal data and on the free movement of such data, and repealing Directive 95/46/EC [2016] OJ L 119 (GDPR). In particular, see art. 22.

43 See e.g. Christoph Busch and Alberto De Franceschi 'Granular Legal Norms: Big Data and the Personalization of Private Law' in Vanessa Mak, Eric Tjong Thin Tai and Anna Berlee (eds.) *Research Handbook on Data Science and Law* (Elgar 2018); Margot E. Kaminski and Gianclaudio Malgieri, 'Algorithmic impact assessments under the GDPR: producing multi-layered explanations' (2020) *International Data Privacy Law*; Mireille Hildebrandt, 'Profiling and the Rule of Law' (2009) 1 *Identity Inf Soc* 64.

44 Sandra Wachter, Brent Mittelstadt and Luciano Floridi 'Why a Right to Explanation of Automated Decision-Making Does Not Exist in the General Data Protection Regulation' (2017) *International Data Privacy Law*; Sandra Wachter and Brent Mittelstadt, 'A Right to Reasonable Inferences: Re-Thinking Data Protection Law in the Age of Big Data and AI' (2019) 2 *Columbia Business Law Review*; Gianclaudio Malgieri and Giovanni Comandé, 'Why a Right to Legibility of Automated Decision-Making Exists in the General Data Protection Regulation' (2017) 7(3) *International Data Privacy Law*; Margot E. Kaminski, 'The Right to Explanation, Explained' (2019) 34(1) *Berkeley Technology Law Journal*, 15; Andrew Selbst and Julia Powles, 'Meaningful Information and the Right to Explanation' (2017) 7(4) *International Data Privacy Law* 233-242; Alessandro Mantelero, ' From Group Privacy to Collective Privacy: Towards a New Dimension of Privacy and Data Protection in the Big Data Era (2017) *Group Privacy* 139-158.

45 Veronica Marotta, Vibhanshu Abhishek and Alessandro Acquisti, 'Online Tracking and Publishers Revenues: An Empirical Analysis' (2019) https://www.semanticscholar.org/paper/Online-Tracking-and-Publishers-Revenues%3A-An-Marotta/bee63f4551c7b6a5a1f07357734a81eab2fec919.

46 See Arlen Moller, Richard Ryan and Edward Deci, 'Self-Determination Theory and Public Policy: Improving the Quality of Consumer Decisions without using Coercion' (2006) 25(1) *Journal of Public Policy & Marketing* 104-116; also Fabrizio Esposito, 'Conceptual foundations for a European Consumer Law and Behavioural Sciences Scholarship' in Hans-W. Micklitz, Anne-Lise Sibony, Fabrizio Esposito (eds.) *Research Handbook in Consumer Law* (Elgar 2018).

they attribute to that set a specific saliency. Hence consumers are deprived of general understanding regarding the state of the market and the behaviour of their peers, which is pivotal for them to develop purchase preferences consciously and autonomously.[47] Furthermore, this effect is exacerbated by the frequent inability of consumers to recognize the factitious nature of what they find online or understand the way profiling algorithms can craft what is offered to them. Frequently, this form of discrimination has been investigated as a form of manipulation, or nudge,[48] and therefore it could be argued prima facie that it actually constitutes an expression of the first form described above. Yet there is a profound difference. Whereas manipulation involves an active (or sometimes malicious) intent to direct consumers towards a certain product or service, the reduction of individual perception of the true state of the market emerges as an inherent consequence of profiling. In this sense, the threat to autonomy also differs from the (previously examined) exploitation of consumer bias, operating as an exogenous effect of pervasive market segmentation rather than as an effect of individual heuristics.

In conclusion, oftentimes – and regardless of the specific kind of discrimination addressed – tailored techniques have been investigated from the common procedural standpoint of explainability. This perspective is focused on how to empower consumers and enable them to inspect and contrast incorrect decisions caused by software arbitrariness in conducting the profiling process, or by errors present in the dataset (this aspect is often traced back to debate regarding the black-box problem)[49] when algorithms are used by private subjects and, especially, public administration.[50] Although a procedural perspective proves to be pivotal in ensuring the effectiveness of protection, an understanding of substantive risks related to the formation of the parties' free will when personalized practices are implemented, in a proactive perspective, is equally (or even more) relevant.

3. A primer on attempts to regulate algorithmic discrimination

As a corollary of the extensive investigation of the potential discriminatory effects embedded in targeting strategies, European experts attempted to identify de iure condito regulatory responses that might prove effective in enhancing consumer protection in the digital environment. In particular, attention was devoted to the role of data protection, antitrust rules, and consumer law. None of these solutions, however, seems conclusive or robust enough to encompass the multifaceted risks that personalized commercial practices entail. Specifically, extensive research has been carried out regarding the capability of the General Data Protection Regulation (GDPR) to provide effective regulation of the data management and processing methods implemented in profiling algorithms.

Privacy and data protection scholars have called for a functional interpretation of GDPR-related user rights (e.g. rights related to individual automated decision-making, explanation, and the right to access) as a tool to disentangle the computerized process and equip data subjects with the concrete ability to infer information regarding the use of their data and its impact on the commercial offerings directed at them. Yet, as was previously mentioned, this perspective focuses primarily on the governance of data processing and acquisition. This approach proves to be inherently incomplete, since important values other than consumer privacy are present and significant.[51] Indeed, even though data protection rules properly regulate the acquisition and processing of users' personal information by data controllers and processors – and, in this sense, operate as an enabling factor for consumer protection[52] – they nevertheless provide only marginal protection for other individual rights and freedoms such as personal autonomy and self-determination. This happens, first and foremost, because the scope of data protection law is limited to personal data and, therefore, personalized practices are not bound to GDPR rules as long as users' data can be anonymized or is non-personal. In addition, the GDPR tackles information asymmetries and privacy risks by empowering consumers regarding which, how, and for what purpose data is acquired and processed; it does not, however, address the systemic effects that profiling likely introduces in terms of individual self-determination and the ability to develop purchase preferences. Protecting the structural state of the market is, indeed, beyond the regulation's scope. Lastly – and acknowledging the fact significant efforts have been made to introduce privacy-by-design solutions to

47 In addition, it has been observed that impairing consumers' sense of autonomy when making choices affects their well-being, diminishing their perception of being in control of their choices. See Quentin André, Ziv Carmon, Klaus Wertenbroch, Alia Crum, Douglas Frank, William Goldstein, Joel Huber, Leaf van Boven, Bernd Weber and Haiyang Yang, 'Consumer Choice and Autonomy in the Age of Artificial Intelligence and Big Data' (2018) 5 *Cust Need and Solut* 28–37.

48 Calo (n 17); Karen Yeung, ''Hypernudge': Big data as a mode of regulation by design' (2018) 20 *Communication and Society* 118-136.

49 Frank Pasquale, The *Black Box Society: The Secret Algorithms That Control Money and Information* (HUP 2015); Sandra Wachter, Brent Mittelstadt and Chris Russel, 'Counterfactual Explanations Without Opening the Black Box: Automated Decisions and the GDPR' (2018) 31(2) *Harvard Journal of Law & Technology*; Matthias Leese, 'The New Profiling: Algorithms, Black Boxes, and the Failure of Anti-Discriminatory Safeguards in the European Union' (2014) 45 *Security Dialogue* 5. On explainability in general, see *inter alia* Frank Pasquale, 'Toward a Fourth Law of Robotics: Preserving Attribution, Responsibility, and Explainability in an Algorithmic Society' (2017) 78(5) *Ohio State Law Journal* 1243-1255; Jack Balkin 'The Three Laws of Robotics in the Age of Big Data' (2017) ibidem 1217-1241; Bryce Goodman and Seth Flaxman, 'European Union regulations on algorithmic decision making and a 'right to explanation'' (2017) 38(3) *AI Magazine* 76–99; Andrew Selbst and Julia Powles, 'Meaningful information and the right to explanation', (2017) 7(4) *International Data Privacy Law* 233–242.

50 As regards the latter aspect, Member State courts are increasingly developing principles that could enhance transparency when using automated systems in executing administrative activities. See e.g. in Italy Consiglio di Stato, judgment of 8 April 2019, n. 2270; Michael W Monterossi, 'Algorithmic Decisions and Transparency: Designing Remedies in View of the Principle of Accountability' (2019) 5(2) *Italian Law Journal* 711-730. More generally, see Hans Micklitz and Przemyslaw Palka, 'Algorithms in the Service of the Civil Society' (2019) 8(1) *Journal of European Consumer and Market Law* 2.

51 See Jabłonowska et al. (n 33). It is not by chance that, in recent years, data protection scholars have begun to reconcile different rights involved in profiling for commercial purposes under the common framework of the right to information self-determination (see, critically, Bert-Jaap Koops, 'The Trouble with European Data Protection Law' (2014) 4(4) *International Data Privacy Law*, 250-261), expanding the basis provided by Art. 8 of the EU Charter of Fundamental Rights and advocating in favour of a wider role for data protection law in informing consumer rights. For a further exploration of the scope and meaning of Art. 8, Case C-40/17 Fashion ID GmBH & Co KG v Verbraucherzentrale NRW eV. [2019] ECLI:EU:C:2019:629. See also Heiko Richter, 'The Power Paradigm in Private Law. Towards a Holistic Regulation of Personal Data', in Mor Bakhoum, Beatriz Conde Gallego, Mark-Oliver Mackenrodt, Gintar Surblyt -Namavi ien (eds.), *Personal Data in Competition, Consumer Protection and Intellectual Property Law* (Springer 2018) 565; Helena Ursic, 'The Failure of Control Rights in the Big Data Era: Does a Holistic Approach Offer a Solution?' (2018) *ibidem*, 55.

52 Manon Oostveen and Kristina Irion, 'The Golden Age of Personal Data: How to Regulate an Enabling Fundamental Right?' in Mor Bakhoum, Beatriz Conde Gallego, Mark-Oliver Mackenrodt, Gintar Surblyt -Namavi ien (eds.) *Personal Data in Competition, Consumer Protection and Intellectual Property Law* (Springer 2018), 8.

make algorithms more responsible[53] – provisions in the GDPR are still heavily reliant on disclosure duties as the main strategy to empower consumers and ensure conscious consent. This occurs both in cases where personal data is collected directly from data subjects and when it is obtained through third parties.[54] Yet, a vast number of empirical studies have warned against the actual efficacy of this tool and have raised doubts regarding the likelihood it could improve decision-making, since consumers systematically tend not to read privacy policies or tend to misunderstand them.[55]

Related to the assumption that the heart of the problem lies in the characteristics of data-driven network architecture, there is also growing interest in competition law as a tool to tackle the distortions caused by personalized practices, in order to promote the establishment of fundamental rights in the European framework. This tendency has developed steadily, along with increasing efforts by public powers to regulate big data companies' ever-expanding exercise of power in digital markets, both in the European Union and abroad. It has been further fostered by recent decisions such as the one involving the German Bundeskartellamt and Facebook between 2019 and 2020.[56] Acknowledging the fact that the conduct of digital platforms is not yet subject to comprehensive and enforceable regulation, competition agencies seem to be increasingly willing to step in and use their enforcement powers to combat new forms of consumer harm[57] and to contrast big tech companies' causal-structural and modal power.[58]

This approach raises concerns as well. One concern is the ability of competition law to adapt its notions as they have been traditionally interpreted (e.g. relevant market, causality, and even harm)[59] to the specifics of the data market has been disputed. Another concern is that the true adequacy of antitrust public enforcement remedies – considering both fines and orders – in directly promoting consumer protection is questionable (and, it could be said, falls beyond the inherent scope of competition law). This is especially the case in light of the open-ended nature displayed by the remedies that have been issued in the abovementioned judgements.[60]

As far as consumer protection law is considered, it should not surprise anyone that the vast majority of scholars have explored the topic of personalized commercial practices by referring to the regulation provided by the Unfair Commercial Practices Directive (hereafter UCPD),[61] especially in light of the innovations proposed in the New Deal for Consumers[62] and the amendments subsequently introduced by the so-called Modernization Directive[63] in the UCPD and in the Directive on Consumer Rights.[64]

Without a doubt, rules prohibiting unfair and, in particular, misleading commercial practices[65] are attractive prima facie solutions in reducing the risks inherent in tailored strategies. Under the UCPD, a commercial practice is qualified as unfair when it is likely to materially distort the economic behaviour of the average consumer through techniques that impair their ability to make informed decisions, causing them to make transactional choices they would not have taken otherwise. In addition, a practice is specifically qualified as misleading if it is likely to deceive the personalized consumer. It is, therefore, not surprising that consumer law scholars have argued in favour of applying these provisions to protect consumers against potential discriminations caused by personalized strategies.[66]

53 Lilian Edwards and Michael Veale, 'Slave to the Algorithm? Why a 'Right to an Explanation' is probably not the remedy you are looking for' (2017) 16 Duke Law & Technology Review; as for contributions analysing the privacy by design principle in general, see Ira Rubinstein 'Regulating Privacy By Design' (2012) 26 *Berkeley Technology Law Journal* 1409.

54 See Arts. 13 and 14 GDPR.

55 See *inter alia* Ian Ayres and Alan Schwartz, 'The No-Reading Problem in Consumer Contract Law' (2015) 66 *Stanford Law Review* 545; Omri Ben-Shahar, 'The Myth of the 'Opportunity to Read' in Contract Law' (2009) 1 European Review of Contract Law; Oren Bar-Gill and Franco Ferrari, 'Informing Consumers about Themselves' (2010) 3 *Erasmus Law Review*, 93.

56 Bundeskartellamt, decision no B6-22/16 of 6 February 2019, Facebook Inc., Menlo Parc, U.S.A., Facebook Ireland Ltd., Dubin, Ireland, Facebook Deutschland GmbH/Verbraucherzentrale Bundesverband e. V., Berlin.; OLG Düsseldorf, Order of 9 January 2015, Az. VI Kart 1/14 (V) - (HRS) juris; Bundesgerichtshof; decision no KVR 69/19 of 23 June 2020. See also the recent request for a preliminary ruling against this decision, submitted to the European Court of Justice by the Bundeskartellamt on 5 March 2021.

57 See Anne C. Witt, 'Excessive Data Collection as Anticompetitive Conduct – The German Facebook Case' (2019) 8 Jean Monnet Working Paper https://jeanmonnetprogram.org/paper/excessive-data-collection-as-anticompetitive-conduct-the-german-facebook-case; Marco Botta and Klaus Wiedermann, 'The Interaction of EU Competition, Consumer, and Data Protection Law in the Digital Economy: The Regulatory Dilemma in the Facebook Odyssey' (2019) 64 *Antitrust Bulletin* 428-446; Justus Haucap, 'Data Protection and Antitrust: New Types of Abuse Cases? An Economist's View in Light of the German Facebook Decision' (2019) *CPI Antitrust Chronicles* https://www.competitionpolicyinternational.com/data-protection-and-antitrust-new-types-of-abuse-cases-an-economists-view-in-light-of-the-german-facebook-decision; Giuseppe Colangelo and Mariateresa Maggiolino, 'Data Protection in Attention Markets: Protecting Privacy through Competition?' (2017) 8 *Journal of European Competition Law & Practice*.

58 Maureen Ohlhausen and Alexander Okuliar, 'Competition, Consumer Protection, and the Right [Approach] to Privacy' (2015) 80 *Antitrust Law Journal* 121.

59 See Petit (n 39); Inge Graef, 'Market Definition and Market Power in Data: The case of Online Platforms' (2015) 38 *World Competition: Law and Economics Review* 4; Simonetta Vezzoso 'Competition Policy in a world of Big Data' in X Olleros and M Zhegu (eds.) *Research Handbook on Digital Transformations* (Cheltenham 2016).

60 Inge Graef 'Blurring Boundaries of Consumer Welfare: How to Create Synergies Between Competition, Consumer and Data Protection Law in Digital Markets' in Mor Bakhoum, Beatriz Conde Gallego, Mark-Oliver Mackenrodt, Gintar Surblyt -Namavi ien (eds.), *Personal Data in Competition, Consumer Protection and Intellectual Property Law* (Springer 2018), 122; Roberto Pardolesi, Roger Van Den Bergh and Fransiska Weber, 'Facebook e i portenti del 'Konditionenmissbrauch'' (2020) 3 *Mercato, concorrenza, regole*.

61 Directive 2005/29/EC of the European Parliament and of the Council of 11 May 2005 concerning unfair business-to-consumer commercial practices in the internal market and amending Council Directive 84/450/EEC, Directives 97/7/EC, 98/27/EC and 2002/65/EC of the European Parliament and of the Council and Regulation (EC) No 2006/2004 of the European Parliament and of the Council [2005] O JEC L 149/22 (UCPD). Regarding the role of the UCPD in tackling personalized practices, see *ex multis* Philipp Hacker, 'Personalized Law and the Behavioral Sciences' in Christoph Busch and Alberto De Franceschi (eds.) *Algorithmic Regulation and Personalized Law. A Handbook* (Hart 2021), 252.

62 Communication from the Commission to the European Parliament, the Council and the European Economic and Social Committee. A new deal for consumers Brussels [2018] COM 183 final.

63 Directive (EU) 2019/2161 of the European Parliament and of the Council of 27 November 2019 amending Council Directive 93/13/EEC and Directives 98/6/EC, 2005/29/EC and 2011/83/EU of the European Parliament and of the Council as regards the better enforcement and modernisation of Union consumer protection rules [2019] OJ L 328/7 (Modernisation Directive).

64 Directive 2011/83/EU of the European Parliament and of the Council of 25 October 2011 on consumer rights, amending Council Directive 93/13/EEC and Directive 1999/44/EC of the European Parliament and of the Council and repealing Council Directive 85/577/EEC and Directive 97/7/EC of the European Parliament and of the Council [2011] OJ L 304, 64–88 (hereafter CRD).

65 See Arts. 5 and 6 UCPD.

66 *Ex multis* Alexandre Streel and Florian Jacques, 'Personalised pricing and EU law' (2019), 30th European Conference of the International Telecommunications Society (ITS):'Towards a Connected and Automated Society', Helsinki, Finland, 16th-19th June, 2019 https://www.econstor.eu/

In spite of the doubtless appeal of regulating tailored practices through the UCPD, this option presents significant limitations as well. First and foremost, personalized strategies are difficult to reconcile with the main categories employed in the UCPD, as they blur the boundaries between (lawful) promotion of products through mere persuasion and (unlawful) manipulation in the assessment of the practice, given the intricacy of defining concepts such as 'unfairness' and 'misleading' in prescriptive terms[67] and the difficulty of reconciling a technique which is inherently based on personalization with normatively determined (and contested) standards such as the concept of 'average consumer'.[68] This is also in consideration of the fact that – even when specific vulnerable groups are present[69] – the UCPD always requires a commercial practice to be defined as deceitful with respect to its targeted group's average member, which must be taken as the benchmark.[70] Such a standard is inherently problematic if we wish to reconcile it with the heterogeneity of behavioral biases present in a population (which are difficult to relate to specific target groups)[71] and with the inherent structure of personalized practices. The aim is to progressively overcome a 'clustered' approach to consumer groups and individualize interaction.[72]

bitstream/10419/205221/1/de-Streel-Jacques.pdf; Federico Galli 'Online Behavioural Advertising and Unfair Manipulation Between the GDPR and the UCPD' in Martin Ebers and Marta Cantero Gamito (eds.) *Algorithmic Governance and Governance of Algorithms. Legal and Ethical Challenges* (Springer 2020), 110-132.

67 Chris Willet, 'Fairness and Consumer Decision Making under the Unfair Commercial Practices Directive' (2010) 33 *Journal of Consumer Policy* 247-273; Mateja Djurovic, *European Law on Unfair Commercial Practices and Contract Law* (Hart 2016).

68 See *ex multis* Hans-W. Micklitz, 'Unfair commercial practices and misleading advertising' in Norbert Reich, Hans-W. Micklitz, Peter Rott and Klaus Tonner (eds.) *European consumer law* (Intersentia 2014) 67-123; Stephen Weatherill 'Who is the 'Average Consumer?' (2009) in Stephen Weatherill and Ulf Bernitz (eds.), *The Regulation of Unfair Commercial Practices under EC Directive 2005/29. New Rules and New Techniques* (Hart 2007) 119; Vanessa Mak 'Standards of Protection: In Search of the 'Average Consumer' of EU Law in the Proposal for a Consumer Rights Directive' (2010) 4 Tisco Working Paper Series on Banking, Finance and Services, 1-16; Cees Van Dam, 'The Average Consumer – a pluriform phenomenon' (2009) 3 *Tijdschrift voor Europees en economisch recht*, 11; Bram Duivenvoorde, *The consumer benchmarks in the Unfair Commercial Practices Directive* (Springer 2015).

69 See Art. 5.3 UCPD.

70 See Art. 5.2 UCPD. Also, Bram Duivenvoorde, 'The Protection of Vulnerable Consumers under the Unfair Commercial Practices Directive' (2013) 2(2) *Journal of European Consumer and Market Law* 69-79.

71 Hacker (n 61), 258.

72 As a consequence of this consideration, a vast debate has arisen in recent years regarding the advisability to promote the creation of personalized standards, using big data analytics and artificial intelligence to tailor each provision to individual needs and characteristics. See Christoph Busch and Alberto De Franceschi, 'Personalization and Granularity of Legal Norms in the Data Economy: A Transatlantic Debate' in Christoph Busch and Alberto De Franceschi (eds.) *Algorithmic Regulation and Personalized Law. A Handbook* (Hart 2021), 3; Tony Casey and Anthony Niblett, 'Self-driving Laws' (2016) 66(4) *University of Toronto Law Journal* 426; Id., 'Framework for the New Personalization of Law' (2019) 86(2) *University of Chicago Law Review* 333-358. This approach is seen to be promising in specific areas of law, such as the drafting of disclosures, see for instance Joasia Luzak, 'Tailor-made Consumer Protection: Personalization's Impact on the Granularity of Consumer Information' in Marcelo Corrales Compagnucci, Helena Haapio, Margaret Hagan and Michael Doherty, *Legal Design: Integrating Business, Design and Legal Thinking with Technology* (Edward Elgar 2021). However, a general claim for legal personalization is generally acknowledged to be a questionable strategy, see Christoph Grigoleit and Philip Maximilian Bender, 'The Law between Generality and Particularity. Chances and Limits of Personalized Law' in Christoph Busch and Alberto De Franceschi (eds.) *Algorithmic Regulation and Personalized Law. A Handbook* (Hart 2021) 132; Marietta Auer, 'Granular Norms and the Concept of Law: A Critique' *ibidem* 137.

The fact that current regulatory interventions to address microtargeting have focused solely on the aspect of price discrimination and price sensitivity[73] seems to further confirm the difficulties that consumer law is facing in dealing with this phenomenon. In addition, the choice of regulating this topic via mandatory information for consumers regarding the existence of personalized prices 'so that they can take into account the potential risks in their purchasing decision' seems to overlook the previously mentioned debate on the shortcomings of disclosure duties in B2C relationships[74] and the fact that transactional decisions by average consumers are determined not only by prices, but more generally by purchasing conditions as a whole.[75]

Furthermore, as was underscored before, problems are not limited to the payment of different prices for the same product amongst profiled consumers. Rather, they extend to risks related to creating a fictional perception regarding the actual presence of different products on the market and making it impossible to observe the behavior of peers, which is regarded as a significant part of the learning process in consumption.[76]

Lastly, a major – and extensively explored[77] - problem in addressing personalized practices via the UCPD is related to private enforcement. In the Directive, no indication is present regarding the appropriate remedy that should be issued after the violation of its provisions. This is an intentional choice, as emerges from Recital 9 of the Directive, which states that the norms in the UCPD operate 'without prejudice to individual actions brought by those who have been harmed by an unfair commercial practice, [...] and without prejudice to Community and national rules on contract law'. Accordingly, Member State governments (and, potentially, courts in individual cases) are required to set rules to foster the Directive's implementation.

Regarding this aspect, which has often been pinpointed as critical in terms of consumer protection,[78] steps forward are currently being taken. The Modernization Directive encourages private enforcement for consumers who are victims of unfair commercial practices by requiring Member States to make proportionate and effective remedies available to them, with specific reference to rights to damages and (if relevant) the unilateral termination of the contract.[79] Yet the actual choice regarding the appropriate remedy and the conditions for its adjudication are still remitted to Member States' national laws,

73 Art. 4 of the Modernization Directive. See also Willem van Boom, Jean-Pierre I. van der Rest, Kees van den Bos & Mark Dechesne, 'Consumers Beware: Online Personalized Pricing in Action! How the Framing of a Mandated Discriminatory Pricing Disclosure Influences Intention to Purchase' (2020) 33 *Soc Just Res* 331–351.

74 See *supra* n 54.

75 Sebastião Barros Vale 'The Omnibus directive and online price personalization: a mere duty to inform?' (2020) 2 *European Journal of Privacy Law & Technologies*.

76 See Aihui Chen, Yaobin Lu and Bing Wang, 'Customers' purchase decision-making process in social commerce: A social learning perspective' (2017) 37(6) *International Journal of Information Management* 627-638; Enrico Moretti, 'Social Learning and Peer Effects in Consumption: Evidence from Movie Sales' (2011) 78(1) *The Review of Economic Studies* 356-393; Markus M. Mobius and Tanya S. Rosenblat, 'Social Learning in Economics' (2014) *Annual Review of Economics* 6 827-847.

77 Hugh Collins, 'Harmonisation by Example: European Laws against Unfair Commercial Practices' (2010) 73(1) *The Modern Law Review* 89-118; Tihamer Toth (ed.), *Unfair Commercial Practices: The Long Road to Harmonized Law Enforcement* (Pázmány 2014).

78 Franziska Weber, 'Abusing Loopholes in the Legal System – Efficiency Considerations of Differentiated Law Enforcement Approaches in Misleading Advertising' (2012) 5(4) *Erasmus Law Review*, 289; Willem van Boom, 'Experiencing Unfair Commercial Practices: An Introduction' (2012) *ibidem* 234.

79 See Recital 16 of the preamble of the Modernization Directive.

and this has significant consequences in terms of the effectiveness of enforcement (especially considering the inner trans-nationality of the digital market). Indeed, normative fragmentation exacerbates the abovementioned problems related to regulating the procedural dimension of anti-discriminatory enforcement. It creates uncertainty for consumers, deterring claims and ultimately curbing access to justice.

4. The 'porous' nature of European private law and the potential of rules on consent

Various solutions could be explored in order to address the shortcomings of the rules concerning tailored commercial practices and, at the same time, exploit the opportunities offered by these developments to produce a better framework for consumers. Regarding this aspect, it is important to stress that some notions and tools in private law can be 'porous'[80] enough to allow for an oriented interpretation that can be functional to regulating personalized strategies. These tools can offer a sufficient margin of appreciation to incorporate emerging findings and be viable instruments in the modernization of consumer protection. As a matter of fact, profiling affects contractual relationships and market exchanges alike. Hence, the interaction between private law and consumer rules is plausible in order to protect the interests of individuals throughout the market experience.[81] In addition, discrimination through tailored offers is a reduction of self-determination that has an impact on consumers' capacity to genuinely develop their free will in contracts, which is an aspect consistently addressed by private law rules.

At the same time, it is common knowledge that the possibility of applying private law rules to the field of consumer protection is not undisputed amongst legal scholars, and that interpretations of the relationship between the two areas vary significantly amongst the legal regimes of the Member States, on the basis of different grounds. Examples include the allegedly different needs and goals pursued by the two bodies of regulation or the diverse conceptual approaches to consumer vulnerabilities they entail.[82]

It is beyond the scope of this work to investigate the general relationship between consumer and private law. Still, it is reasonable to defend the view that – despite their substantial differences and considering the role of European private law – it is not necessarily the case that contract law and consumer protection have to be dedicated to pursuing completely different goals. The two sets of regulations are meant to promote free and frequent exchanges by protecting both parties' genuine consent in order to, ultimately, make the most of the rationality of operators and to respect the fundamentals of a market economy. Furthermore, as far as the concomitant value dimension of contracts is concerned, they both pursue egalitarian goals, in the sense that they seek to balance disparities amongst unequal parties that might otherwise produce an unfair result for vulnerable persons, and to harmonize the autonomy of the parties along the lines of policies of social and distributive justice.[83]

In light of this teleological symmetry, it is reasonable to believe it is possible to apply general remedies (from broader branches of regulation such as private and contract law) to consumer law, as integrative resources. This perspective is, furthermore, consistent with the hierarchical relationship between lex generalis and lex specialis[84] – just as it is between European private law (considered an autonomous field, separate from that of Member States) and consumer law. Accordingly, it is legitimate for contract law to operate as an ancillary resource when provisions developed within consumer law do not yield effective answers to commercial strategies based on technological strategies that consumer law does not (yet) adequately address, as is the case in the field of personalized advertising.

Amongst private law rules, in particular, provisions on defective consent might be a viable regulatory solution[85] and a tool to promote social justice in personalized interactions, by contributing to a broader framework for the assessment and regulation of targeted services pursuant to substantive fairness in contractual relationships. From a general perspective, to evaluate whether a contract should be avoided, rules on defective consent are designed to take into consideration different situations affecting the formation of a party's genuine assent to the conclusion of a contract. According to the structure depicted in the main set of rules for international and European contexts, an initial hypothesis (mistake) occurs every time a party shows an incorrect understanding of the content of a contract as a result of an erroneous analysis of the agreement and its provisions, based on her own belief. Moreover, a different hypothesis (misrepresentation) occurs if the counterparty – even acting in good faith – made or caused this mistake, or knew or ought to have known of the mistake, and willfully left the mistaken party in error, and the counterparty knew that the mistaken party, had they known the truth, would not have entered into the contract or would have done so only on fundamentally different terms. Lastly, a third hypothesis (fraud) arises in situations in which the provision of consent is determined by an intentional false statement of facts by the counterparty, meant to deceive the contractor.

This general structure is not unambiguous amongst jurisdictions, with doctrines heterogeneously construing the three concepts.[86]

80 Genevieve Helleringer and Anne-Lise Sibony 'European Consumer Protection Through The Behavioral Lens' (2017) 23(3) *Columbia Journal Of European Law*.

81 See Domurath (n 2) 88.

82 Carmelita Camardi, 'Pratiche commerciali scorrette e invalidità' (2010) 6 *Obbl contr* 408.

83 See Chantal Mak, *Fundamental Rights in European Contract Law: A Comparison of the Impact of Fundamental Rights on Contractual Relationships in Germany, the Netherlands, Italy and England* (Kluwer 2008), 50.

84 With specific regard to the relationship between EU private law and consumer protection law, see Vanessa Mak, 'The Consumer in European Regulatory Private Law', in Dorota Leczykiewicz and Stephen Weatherill (eds.), *The Image of the Consumer in EU Law: Legislation, Free Movement and Competition Law* (Hart 2016), 381-400. See also Dorota Leczykiewicz and Stephen Weatherill (eds.), *The Involvement of EU Law in Private Law Relationships* (Hart 2013); Hans-W. Micklitz, 'Unfair commercial practices and European private law', in Christian Twigg-Flesner (ed.), *The Cambridge Companion to European Union Private Law* (CUP 2010), 229-242.

85 Fabrizio Cafaggi and Horatia Muir Watt, *The Regulatory Function of European Private Law* (Elgar 2009); Hans-W. Micklitz, 'The Visible Hand of European Regulatory Private Law – The Transformation of European Private Law from Autonomy to Functionalism in Competition and Regulation' (2009) 28 Yearbook of European Law 3-59; Hans-W. Micklitz, *The Politics of Justice in European Private Law: Social Justice, Access Justice, Societal Justice* (CUP 2018).

86 For example, according to § 119 of the German *Bürgerliches Gesetzbuch* (BGB), mistakes need not to be known by the counterparty – with this element constituting the essential divide between mistakes and deceit *ex* § 122(1) BGB – and, therefore, it might entitle them to receive reliance damages from the party avoiding the contract. Differently, in the Italian legal system, a party's mistakes must be recognizable by the counterparty in order to justify the avoidance of the contract; see Art. 1431 of the Italian Civil Code (*Codice Civile*). The distinction between misrepresentation and mistake has been subject to prominent debate in common law jurisdictions as well: the 1967 Misrepresentation Act distinguishes between fraudulent, negligent, and innocent actionable misrepresentation as basis for recission, whereas the doctrine of mistake (which can be common, mutual, and unilateral) developed mostly through case law (*ex multis Bell*

Moreover, a neat distinction between mistake, misrepresentation and fraud is not always present in regulations.[87] Still, in spite of different names and some discrepancies in their configurations, rules on defective consent are present – and follow similar structures – in the vast majority of Member States, having their conceptual common core in the Roman tradition;[88] all European systems acknowledge the view that an expression of will might arise from a (self- or hetero-determined) misrepresentation of the characteristics of the agreement.[89] As a further confirmation, rules on defective consent are present both in the Principles of European Contract Law (PECL) regulating the means to avoid a contract due to a mistake[90] or fraud generated by the counterparty,[91] and in the Draft Common Frame of Reference (DCFR) within provisions related to fraud and good faith in fair dealing.[92]

Considering these aspects, academics have already suggested applying these rules as supplementary resources to tackle other shortcomings in the regulation of commercial practices that were investigated in previous years, such as the exploitation of consumers' cognitive biases.[93] Building on this experience, the application of rules on defective consent to tailored commercial techniques could foster an enhancement of the level of consumer protection in the digital environment and overcome the various critical aspects of the above-mentioned regulations.

First of all, and similar to the GDPR, rules on defective consent arise from the common ground of protecting consumers' information self-determination, while they also exhibit a wider and more flexible scope. On the one hand, they are suitable for regulating not only the acquisition and processing of data that is functional to personalized advertising and profiling, but also the entire B2C interaction. On the other hand, they are disentangled from the inner weaknesses of information duties as a means of generating genuine consent. In addition to shortcomings related to the no-reading problem, mandated disclosures are, as a matter of fact, circumscribed in many aspects. Namely, they must be identified ex ante and they usually grant victims the right to ask for compensation only, without affecting the validity of the contract concluded. Lastly, information duties are inherently fragmentary, meaning that the same information can be framed from different perspectives. For example, mandated notice regarding the performance of profiling strategies (and even personalized pricing) can be represented as a process conducted in a client's best interest, in order to find the most suitable product, while the consequent reduction of choice is not mentioned.

Against this backdrop, rules on defective consent provide an ex-post tool for judicial scrutiny, as they devote specific attention to the interpretation of the parties' behaviour – and implemented strategies – throughout the whole bargaining process, including the pre-negotiation phase. In this way, the assessment of unlawfulness conducted in accordance with the rules on misrepresentation and fraud can consider the entirety of elements that contributed to the formation of the contract. Thus, when a party's conduct artificially affects the understanding that the counterparty has regarding the characteristics or the functioning of a product or a service (e.g. by extremely narrowing the selection of products offered, so as to induce a state of almost complete 'blindness' in the consumer regarding the state of the market), these rules can provide solid ground for the elimination of the harmful effects of the contract.

Significant advantages are also present when interaction with regulation on unfair commercial practices is considered. Once again, both set of rules start from a common conceptual ground, namely the unfair modification of one party's will. Yet provisions on defective consent do not require the consumer to take (virtually or in practice) a transactional decision that they would not have taken otherwise, as this is explicitly mentioned as an essential element in both the wordings of Art. 6 (on misleading actions) and Art. 7 (on misleading omissions) of the UCPD.

On the contrary, rules on defective consent do not require a strict causal link between the use of a discriminatory strategy and the decision to conclude a contract,[94] as they are able to regulate both essential and non-essential mistakes and fraud, as long as these lead to a modification of the agreement's conditions.[95] In addition, these rules are not bound to the rather problematic average consumer benchmark either, which allows courts to perform ex personae scrutiny of each case at stake; consequently, provisions regulating defective consent do not lead to a conclusive statement regarding the tailored practice in se, but rather to the performance of individually segmented evaluations, which are both consistent with the inner characteristics of profiling practices (i.e. their granularization and diversification amongst consumers) and functional to balancing the potentials and shortcomings that these strategies possess.

A reconsideration of the role played by consent rules in the regulation of microtargeting against relying on unfair commercial practices regulation alone is also advisable, when considering, that the latter mainly focuses on the collective protection of consumers at a macroeconomic level.[96] Yet, the sophistication of commercial relationships in the digital environment and the granularization of B2C interaction make it difficult, for a court (or a supervisory authority) to express

v Lever Bros Ltd [1932] AC 161 and *Great Peace Shipping Ltd v Tsavliris Salvage International*) Ltd [2003] QB 679) to identify the requirements for rescission or rectification of the concluded contract. See also Patrick Atiyah and Francis Bennion, 'Mistake in the Construction of Contracts' (1961) 24 Modern Law Review 421; Catharine MacMillian, *Mistakes in Contract Law* (Bloomsbury 2012); John Cartwright, *Misrepresentation, Mistake and Non-disclosure* (Sweet & Maxwell 2012).

87 As it will be observed shortly - the PECL does not rigidly distinguish between mistake and misrepresentation, encompassing them both under Art. 4:103.

88 See Martin Jose Schermaier, 'Mistake, misrepresentation and precontractual duties to inform: the civil law tradition' in Ruth Sefton-Green (ed.), *Mistake, Fraud and Duties to Inform in European Contract Law* (CUP 2005), 39-64.

89 John Cartwright, 'Defects in Consent Contract Law', in Arthur Hartkamp, Martijn Hesselink, Ewoud Hondius, Chantal Mak and Edgar du Perron (eds) *Towards a European Contract Code* (Kluwer 2011), 537.

90 Art. 4:103 PECL.

91 Art. 4:107 PECL.

92 Respectively Art. II.-7:205(1) and Art. II.-7:205(3) DCFR.

93 See Francesco Paolo Zatti, 'Fraud and Misleading Commercial Practices: Modernising the Law of Defects in Consent' (2016) 12(4) European Review of Contract Law 307-334; Jan Trzaskowski, 'Behavioural Economics, Neuroscience, and the Unfair Commercial Practices Directive' (2011) 34(3) Journal of Consumer Policy 377-392; Eleni Tzoulia, 'Imprints of behavioural research in EU consumer protection legislation: the 'average consumer test' in the Unfair Commercial Practices Directive' (2017) *Tijdschrift voor Consumentenrecht en handelspraktijken* 258.

94 See Marco Loos, 'The modernization of European Consumer Law (continued): More meat on the bone after all' (2019) *Amsterdam Law School Legal Studies Research Paper No 2019-32* 3.

95 See e.g. Art. 4:103(1)(b) PECL.

96 Thomas Wilhelmsson, 'Scope of the Directive', in Geraint Howells, Hans-W. Micklitz and Thomas Wilhelmsson (eds.), *European Fair Trading Law. The Unfair Commercial Practices Directive* (Aldershot 2006), 51; Anna Genovese, 'Ruolo dei divieti di pratiche commerciali scorrette e dei divieti antitrust nella protezione (diretta e indiretta della libertà di scelta) del consumatore' (2008) Annali italiani del diritto d'autore della cultura e dello spettacolo 297, 302.

an evaluation on a commercial practice (per se) on a general level, as the UCPD requires. On the contrary, judicial scrutiny conducted through the lens of defective consent can operate as a second-degree evaluation to enrich and correct the outcome of the first-level interpretation under consumer law and allow for a re-assessment based on the specific characteristics of the tailored interaction considered.

This way, rules on defective consent can contribute to broadening the scope of market regulation around justice and substantive efficiency goals, without precluding – when advisable – the direct application of the UCPD. The interaction between both regulatory matters can ameliorate the market process by promoting unhindered decisions, with consumer law working on a broad scale and contract law in the individual case.

While it cannot be claimed that the application of defective consent rules radically erases all incentives for companies to engage in discrimination – these provisions being primarily targeted at enhancing autonomous (consumer) choices – they would nevertheless introduce an additional granularized dimension of scrutiny, which is absent in the UCPD approach, and this might prove to be desirable in reacting to practices that are differentiated on an individual basis. As regards the general provision on unfair commercial practices,[97] private law rules on consent will likely provide more flexibility since they do not require the behaviour to be contrary to the requirements of professional diligence, since it is difficult to break this condition down into specific obligations and standards (whether in terms of implementation or of auditing) when automated processes are considered.[98] Lastly, and underscoring a significant difference from the UCPD, provisions on defective consent provide a certain remedy – avoidance of the contract – as a consequence of violations, which is suitable for protecting consumers and, at the same time, exercising proper deterrence for professionals (especially when coupled with the awarding of compensation for damages for culpa in contrahendo).

On the basis of the characteristics of avoidance, when a contract is vitiated for defective consent as a result of a tailored practice, two alternatives are set for the victim: if the conduct of the professional affected on an aspect of the agreement, which is not necessary (in the eye of the counterparty) for the contract to properly operate, then she will be able to keep the contract in force and ask for compensation based on the professional's culpa in contrahendo.

On the contrary, if the outcome of the exploitation relates to an element that was deemed essential for the conclusion of the contract the party might ask the judge to render the whole contract null and void, then seek damages for its non-conclusion.

This framework of choices that the consumer has at her disposal ultimately shapes a remedy that is, at the same time, flexible and functionally respondent to her specific needs and interests.

Yet, the counterparty's behavior will always be punished - even if its amount will vary depending on the concrete choice of the consumer: when the contract remains in force, the quantum debeatur is quantified considering the worse conditions that the party suffered due to the unfair use of tailored practices;[99] if the contract is declared null as a whole, then the party is instead entitled to be compensated for the conclusion of an invalid agreement.

In summary, rules on defective consent share conceptual ground with the main existing regulatory solutions that were introduced to tackle risks arising from tailored commercial practices. In addition, they overcome some of the current limits that each of them presents and therefore provide a potentially more effective resource for dealing with the phenomenon. Nonetheless, tensions existing between national and European principles of contract law – like those between the different facets embodied in each Member State's rules on defective consent – further epitomize the incompleteness of the system.

Recurring to defective consent rules can raise some points of criticism: it might be argued, for example, that under contract law rules individual consumers would be devoid of enough incentives to pursue protection in court, considering the high risks involved in litigation, the rules regarding the burden of proof, and the fact that the potential benefits may not outweigh its cost.

While acknowledging that, in general, the lack of incentives to act in court constitutes a major concern of the private enforcement system overall – which is found in consumer law as well[100] - it cannot be prima facie excluded that the economic interests linked to the contract may nevertheless persuade the individual to enter in a proceeding. In addition, even being subject to a demanding burden of proof, prior judgments ordering an injunction or a penalty might be useful in alleviating the burden of proof regarding the existence of a fraud or an alteration of consent: in recent years, Member States' jurisdictions held that a public authority's decision might constitute a 'privileged evidence' with regards to a violation of private law rules.[101] Furthermore, this approach is consistent with regulatory initiatives which took place in other areas – such as competition law – regarding follow-on actions;[102] transposing this orientation on the case of defective consent might, therefore, offer a good basis to those individuals who are willing to pursue the avoidance of their contract as a private law remedy.

Lastly, it might be contended that contract law is based on freedom of contract, and therefore should not consider power imbalances. With regards to this aspect, it might be first observed that the understanding of contract law has undergone significant changes in recent years, which are leading to a crescent consensus on the idea of the Materializierung of contract law, taking into account the different bargaining power between the contracting parties and the condition of asymmetric information.[103] According to this perspective, power imbalances would play a significant role in the analysis of defective consent rules as well. Secondly, in the case of tailored commercial practices, the reduction of individuals' autonomy does not (directly) stem from the

97 Art. 5 UCPD.

98 See Sandra Wachter et al (n 28). Some attempts are, nevertheless, present: Nicholas Diakopoulos 'Algorithmic Accountability: the investigation of Black Boxes' (2014) Tow Center for Digital Journalism https://academiccommons.columbia.edu/doi/10.7916/D8ZK5TW2; Dillon Reisman, Jason Schultz, Kate Crawford and Meredith Whittaker, 'Algorithmic Impact Assessments: a practical framework for public agency accountability' (2018) AiNow Institute https://ainowinstitute.org/aiareport2018.pdf; Ebers (n 28) 76.

99 In order to perform this operation, a valid proxy could be represented e.g. by offers made by the same operator to other clients.

100 Franziska Weber, *The Law and Economics of Enforcing European Consumer Law* (Aldershot 2014), 45–52.

101 See the Italian Cass civ, 13 February 2009, Nr 3640 (2010) Il Foro italiano 1901. See also Francesco Paolo Patti, 'Fraud and Misleading Commercial Practices: Modernising the Law of Defects in Consent' (2016) 4 *European Review of Contract Law* 318.

102 For an overview see Pier Luigi Parcu, Giorgio Monti and Marco Botta (eds.) *Private Enforcement of EU Competition Law. The Impact of the Damages Directive* (Cheltenham 2018).

103 Jürgen Basedow, 'Freedom of Contract in the European Union' (2008) 16 *European Review of Private Law* 905.

asymmetry of bargain power between her and the counterparty but, rather, from the consumers' inability to understand the determinants behind the offer presented to her.

5. Concluding remarks

The analysis conducted in this research has shown that rules on fraud and misrepresentation might offer a sufficient margin of appreciation to incorporate emerging findings on personalized practices, and to operate as viable instruments for the modernization of consumer protection in the absence of a form of dedicated regulation. Still, improvements are advisable in order for the system to be optimized. Despite the indications provided by the Principles on European Contract Law and the Draft Common Frame of Reference, and in light of the formal independence of Member States' private law and the (minor, but nevertheless still existing) differences amongst different national rules on defective consent, harmonization is undoubtedly desirable.

While it is beyond the scope of this article to argue extensively in favour of a normative unification of private law in the European framework, it is nevertheless worth observing that the attempt to formulate a uniform set of rules - within the broader conceptual lens of the 'constitutionalization' of private law[104] - has long been identified as a necessary step towards achieving social justice in private relations,[105] and that consumer law has played a pivotal role in stimulating this debate since its earliest days.[106] In addition, and in spite of the difficulties that this process has encountered in recent times, the role of private law as a transformative and conceptually unifying framework has been further stressed by the regulatory uncertainties presented by digital innovations, with a major focus on the transnational dimension of online platforms and commercial practices.[107]

In this context, private law is supposed to operate as the synthesis of the heterogenous experiences of Member States and supervisory authorities filtered through the lens of the fundamental principles that animate the whole European framework and that play a central role in determining the content of regulatory measures. Amongst these principles, the preservation of consumer consent and will (including their perception of the overall existence and characteristics of different products on the market) constitute a necessary condition for the genuine development of the digital environment.

In the absence of a (desirable) stringent harmonization of private law in the European framework, and in light of the (inevitable) shortcomings currently presented by existing regulations (in particular GDPR and UCPD) in addressing high-tech marketing strategies based on personalization, rules on defective consent could provide a valid ad interim solution to attenuate, integrate and correct the possible adverse and discriminatory effects of these techniques, while at the same time preserving the benefits they introduce to the market ecosystem.

Acknowledgements

The author wishes to thank Chantal Mak, Roberto Pardolesi, all the members of the Amsterdam Centre for Transformative Private Law (ACT), the participants in the Maastricht University 'Researching Discrimination in E-Commerce and Online Advertising' webinar and the reviewer for helpful comments on previous drafts of the paper. Any errors that remain are the responsibility of the author. The article was developed as part of the Marie-Sklodowska Curie Project 'FairPersonalization' (MSCA-IF-2019-897310) with the support of the European Commission

104 Hugh Collins 'The Constitutionalization of European Private Law as a Path to Social Justice?' in Hans-W. Micklitz (ed.) *The Many Concepts of Social Justice in European Private Law* (Edward Elgar 2011); See also Jan M. Smits, 'Convergence of Private Law in Europe: Towards a New Ius Commune?' in Esin Örücü and David Nelken (eds.) *Comparative Law: A Handbook* (Hart 2007), 219-240; Martijn W. Hesselink 'The New European Legal Culture' in Martijn W. Hesselink (ed.) *The New European Private Law: Essays On The Future Of Private Law In Europe* (Kluwer 2002), 11-75; Id, 'The General Principles of Civil Law: Their Nature, Roles and Legitimacy' in Dorota Leczykiewicz and Stephen Weatheril (eds.) *The Involvement of EU Law in Private Law Relationships* (Hart 2013), 131-180; Jan M. Smits, 'The Principles of European Contract Law and the Harmonization of Private Law in Europe', in Antoni Vaquer (ed.) *La Tercera Parte De Los Principios De Derecho Contractual Europeo* (Tirant 2005), 567-590.

105 Jürgen Basedow, ‹Codification of Private Law in the European Union: The Making of a Hybrid› (2001) 9(1) *European Review of Private Law* 35–49; Ruth Sefton-Green, 'Social justice and European identity in European contract law' (2006) 2 ERCL 275, 277; Study Group on Social Justice in European Private Law, 'Social Justice in European Contract Law: A Manifesto' (2004) 10(6) European Law Journal 653-674; Chantal Mak, 'Europe-building through private law. Lessons from Constitutional Theory' (2012) 3 *European Review of Competition Law* 326-341; Giovanni Comandé, Gert Brüggemeier and Aurelia Colombi Ciacchi (eds.) *Fundamental Rights and Private Law in the European Union* (CUP) 2010.

106 Albertina Albors-Llorens, 'Consumer Law, Competition Law and the Europeanization of Private Law', in Fabrizio Cafaggi (ed.), *The Institutional Framework of European Private Law* (OUP 1993).

107 *Ex multis* Stefan Grundmann (ed.), *European Contract Law in the Digital Age* (Intersentia 2018); Marija Bartl, 'Socio-Economic Imaginaries and European Private Law' in Poul F. Kjaer (ed.) *The Law of Political Economy: Transformations in the Function of Law* (CUP 2020) 228-253; Hans-W. Micklitz, 'The Transformative Politics of European Private Law' *ibidem* 205-227; Natali Helberger, Lucie Guibault, Marco Loos, Chantal Mak, Lodewijk Pessers and Bart Van der Sloot, *Digital Consumers and the Law: Towards a Cohesive European Framework* (Kluwer 2012); Chantal Mak, 'Fundamental Rights and the European Regulation of iConsumer Contracts' (2008) *Journal of Consumer Policy*, 425-439.

07

Special Issue: Should Data Drive Private Law

Technology and Regulation

Talking at Cross Purposes?

A computational analysis of The debate on informational duties in the digital services and the digital markets acts

Fabiana Di Porto*, Tatjana Grote**, Gabriele Volpi***, Riccardo Invernizzi****

Digital Services Act, Digital Markets Act, Big Platforms, Computational Analysis, Transparency duties

fabiana.diporto@unisalento.it

n.t.grote@lse.ac.uk

gabrielevolpi@me.com

riccardo.invernizzi03@universitadipavia.it

Since the opaqueness of algorithms used on online platforms opens the door to discriminatory and anti-competitive behaviour, increasing transparency has become a key objective of lawmakers. Leveraging the analytical power of Natural Language Processing, this paper investigates whether key terms related to transparency in digital markets were used in the same way by different stakeholders in the consultation on the EU Commission's DSA and DMA proposals. We find significant differences in the employment of terms like 'simple' or 'meaningful' in the position papers that informed the drafting of the proposals. These findings challenge the common assumption that phrases like 'precise information' are used the same way by those implementing transparency obligations and might partially explain why they frequently remain ineffective.

1. Introduction

When EU Executive Vice-President Margarethe Vestager presented the latest Commission proposals on digital platforms, the Digital Markets Act (DMA) and the Digital Services Act (DSA),[1] she compared them to the invention of the traffic light, which was created in response to the rapidly increasing importance of the car. She concluded that 'just like back then, ... now we have such an increase in the online traffic that we need to make rules that put order in the chaos'.[2]

This twin-proposal suggests many new rules for digital intermediary services and online platforms.[3] With the DSA and DMA, the Commission closes a period during which stakeholders (and doctrine)[4] have been harshly discussing new ex ante rules for digital markets, both from a consumer protection and a competition law perspective.[5]

1 European Commission, Proposal for a Regulation of the European Parliament and of the Council on a Single Market For Digital Services (Digital Services Act) and amending Directive 2000/31/EC (COM(2020)825), 15 December 2020 [hereinafter Digital Services Act, DSA]; European Commission, Proposal for a Regulation of the European Parliament and the Council on contestable and fair markets in the digital sector (Digital Markets Act) (COM(2020) 842), 15 December 2020 [hereinafter Digital Markets Act, DMA].

* Fabiana Di Porto is Professor of Law and technology at the University of Salento, Italy and Contract Professor of Innovation Law and Regulation at LUISS, Rome. She is Head of the Algorithmic Disclosure Regulation PRIN-funded Research Project 2019-22 (www.lawandtechnology.it, grant no 2017BAPSXF).We are thankful to Prof. Julian Nyarko from Stanford University for providing some of the data we used in the empirical part and to a number of other people for insightful discussions: Dr Roberto Marseglia, Professors Daniela Piana and Giuseppe Italiano and the anonymous referees. All mistakes remain our own.

** Tatjana Grote has a BA in International Relations from the Technical University of Dresden and is currently an LLM Candidate at the London School of Economics and Political Science as well as a member of the PRIN Algorithmic Disclosure Regulation project 2019-22.

*** Gabriele Volpi has a msc Theoretical physics from the University of Pavia, Manager at DAAT consulting and Member of the PRIN Algorithmic Disclosure Regulation project 2019-22.

**** Riccardo Invernizzi has a bsc in Mathematics - with honors - from the University of Pavia, junior analyst at DAAT consulting and Member of the PRIN Algorithmic Disclosure Regulation project 2019-22.

Received 3 Dec 2020, Accepted 5 May 2021, Published 11 Feb 2022.

2 European Commission, Statement by Executive Vice-President Vestager on the Commission proposal on new rules for digital platforms, 15 December 2020, https://ec.europa.eu/commission/presscorner/detail/en/STATEMENT_20_2450 (accessed 15 February 2021).

3 There is no perfect alignment in the definition of platform services in the DSA and DMA. In the DSA, the widest concept is that of online '*intermediary service*', which covers all services within the scope of Art. 1(3), including 'online platforms' (providing hosting services) under the meaning of Art. 2(1)h DSA. In the DMA, the widest category is that of '*core online platform*'. Art. 2(2) 'online intermediation services' are one service type among the many 'core platform services' (together with e.g., cloud services, social networks, videosharing platforms). Some '*core online platforms*', then, may be designated as 'gatekeepers' (DMA, Art. 3) if they (a) have a significant impact on the internal market, (b) serve as a gateway between business and end-users, (c) enjoy an entrenched and durable position. The requisites are presumed to exist: (i) if the 'core platform service' was provided in at least 3 MS and given thresholds of average market capitalization are overcome; (ii) the core platform has more than 45 million monthly active end-users plus 10.000 business users; (iii) the thresholds in point (ii) were met in each of the last three financial years. (Art. 3, DMA).
Hence, for the sake of parallel applicability of the DSA and DMA transparency rules, not every (core) very large platform is a gatekeeper, but it is likely that every gatekeeper will also be a very large (core) online platform (see Art 3(2)b DMA).

4 P Ibáñez Colomo, Whatever Happened to the 'More Economics-Based Approach'?, *Journal of European Competition Law & Practice* (2020) 11, 9, 473–74, (discussing the shift from the so called 'more economic approach' to the growing demand for ex ante intervention against big digital platforms in the European legal community).

5 For challenges related to competition law, see e.g., A Ezrachi & M Stucke,

Although the two proposals differ in scope and focus,[6] both reveal that one key instrument the Commission relies upon in 'ordering' chaotic traffic in digital markets is informational duties (inclusive of both transparency and disclosure obligations).[7]

This is surprising and unsurprising at the same time. According to the standard narrative, informational duties play a central role in the realm of consumer protection[8] and serve to rebalance unequal bargaining power in trade relationships.[9] And digital markets would be no exception.[10]

On the other hand, the very utility of informational duties has been systematically questioned.[11] Overall, such duties seem to have more of a symbolic (*rectius*, political) value rather than true utility.[12] In the digital realm, many argue that extra-long disclaimers and hard-to-read terms of contract would be useless, or sometimes run counter consumers empowerment.[13] A similar argument is made for platform-to-business relations, where information duties are often considered insufficient to mitigate unequal bargaining power.[14]

This paper aims to investigate *why*, despite the long-lasting scholarly debate about their limited effectiveness, and overwhelming evidence supporting it, the DSA and DMA rely heavily on disclosure.[15] More specifically, we investigate what are the possible *sources* of ineffectiveness.

There have been many attempts to do that, the behavioral literature on disclosure being the most relevant in two regards. On one side, it has provided empirical evidence of the impact of informational arrangements[16] adopted by big digital platforms by measuring how much they affect the behavior of consumers. On the other, it has accounted for the effectiveness of disclosure duties by measuring how many consumers like or dislike them.[17] However, these studies take the legal duty as a given, an external variable. On the contrary, we contend that much can be said about their origin and the process through which this duty is formed.

Therefore, we propose to leverage the power of computational tools, among which Natural Language Processing (NLP) and Machine Learning (ML) techniques: by linguistically analyzing the debate that preceded the adoption of these duties, our empirical study suggests searching for possible sources of failure in the feedback documents to the consultation, that were input to these rules.

Our contribution innovates in several regards. First, our methodology is not effects-based, in the sense that to assess the efficacy of transparency duties, it does not look at the impact on nor the perceptions of those who receive the information, being this input context-specific. We rather analyze the *wording* that conflated the debate around the provisions establishing informational duties of

Virtual competition: the promise and perils of the algorithm-driven economy (Harvard University Press 2016), and P Marsden & R Podszun, estoring Balance to Digital Competition – Sensible Rules, Effective Enforcement, (Konrad-Adenauer-Stiftung 2020), 1-87. On consumer protection and its relation to data protection and competition law, see W Kerber, Digital markets, data, and privacy: competition law, consumer law and data protection, *Journal of Intellectual Property Law & Practice* (2016) 11(11), 856-866.

6 Both the DMA and DSA take a resolute stance, through ex ante regulation, against the big platforms. However, the DSA aims primarily to '*ensur[e] a safe and accountable environment*' by applying asymmetric ex ante rules to online digital platforms, according to two parameters: the company's role (i. intermediary services, ii. hosting services, iii. online platforms), and size (a. large online platforms and b. very large platforms i.e., those reaching more than 45 million consumers, which will have to comply with special rules). The DSA imposes obligations on transparency, illegal content, and accountability requirements. Therefore, it addresses negative externalities and asymmetric information. On the other hand, the DMA's goal is to '*ensur[e] fair and open digital markets*' by applying asymmetric rules against large online platforms designated as '*gatekeepers*', which are addressed with a list of does and don'ts. Taken together, they can be read as an ex ante toolbox, made of a mix of competition and consumer protection rules. While the DSA amends the e-commerce directive (2000/31/EC), the DMA centers around concerns and seeks to complement EU competition rules (mostly Art 101, 102 TFEU). Finally, the DSA applies to all 'intermediary services' (Art 1), while the scope of the latter is limited to 'core platform services' offered by 'gatekeepers' as defined in Art 3 DMA.

7 We use disclosure, transparency and informational duties interchangeably as what is relevant to the analysis is the way the terms related to the provision of information are used by the stakeholders. However, we acknowledge that there are duties owed to users and those to public authorities; and that information may well be provided for purposes of public or private disclosure, or for reasons of investigations. A taxonomy of transparency and disclosure duties is nonetheless provided for in Table 1 in the Appendix, to which reference is made in the legal analysis of Section 2.3 *below*.

8 European Parliament resolution of 20 October 2020 with recommendations to the Commission on the Digital Services Act: Improving the functioning of the Single Market (2020/2018(INL)), 20 October 2020, 12 (no. 31, 32).

9 See e.g., EA Posner, ProCD v. Zeidenberg and Cognitive Overload in Contractual Bargaining. *University of Chicago Law Review* , E. A. (2010) 77(4), 1181-1194.

10 Algorithm Watch (2020), Governing Platforms – Final Recommendations, available at https://algorithmwatch.org/wp-content/uploads/2020/10/Governing-Platforms_DSA-Recommendations.pdf (accessed 17 February 2021), 1.

11 See e.g., O Ben-Shahar & CE Schneider, Coping with the Failure of Mandated Disclosure. *Jerusalem Review of Legal Studies* (2015) *11*(1), 83–93; F Marotta-Wurgler, Even More Than You Wanted to Know About the Failures of Disclosure. *Jerusalem Review of Legal Studies* F. (2015) *11*(1), 63–74. E Zamir, & D Teichman, *Behavioral Law and Economics*. (Oxford University Press 2018), 171-177; F Di Porto, & M Maggiolino, Algorithmic Information Disclosure by Regulators and Competition Authorities. *Global Jurist,* (2019). 19(2), 11; E. Bardach & RA Kagan, *Going by the book: The problem of regulatory unreasonableness.* (Temple University Press 1982), 249-256; A Prat, The Wrong Kind of Transparency. *American Economic Review*, (2005) 95(3), 862.

12 Di Porto & Maggiolino (n 11) 14.

13 SK Ripken,The Dangers and Drawbacks of the Disclosure Antidote: Toward a More Substantive Approach to Securities Regulation. *Baylor Law Review* (2006) 58(1), 160.

14 Marsden & Podszun (n 5), 18; F Di Porto & M Zuppetta, Co-Regulating Algorithmic Disclosure for Digital Platforms, *Policy and Society* (2020) 0(0), 3-4; C Busch, Crowdsourcing, Consumer Confidence: How to Regulate Online Rating and Review Systems in the Collaborative Economy. In C Economy & A De Franceschi (Eds.), *European Contract Law and The Digital Single Market: The Implications of The Digital Revolution*, 223. (Intersentia 2016).

15 See M Sentfleben & C Angelopoulos, The Odyssey of the Prohibition on General Monitoring Obligation on the Way to the Digital Services Act: Between Article 15 of the e-Commerce Directive and Article 17 of the Directive on Copyright in the Digital Single Market, available at: https://papers.ssrn.com/sol3/papers.cfm?abstract_id=3717022 (accessed 23 April 2021) and G Frosio (2020). Taking Fundamental Rights Seriously in the Digital Services Act's Platform Liability Regime, available at: https://papers.ssrn.com/sol3/papers.cfm?abstract_id=3747756, discussing transparency duties in the DSA. For an analysis of disclosure remedies in the DMA, see Ibáñez Colomo P (2021). The Draft Digital Markets Act: A Legal and Institutional Analysis, available at: https://papers.ssrn.com/sol3/papers.cfm?abstract_id=3790276 (accessed 23 April 2021).

16 See e.g., J Luguri & L Strahilevitz, (2021). Shining a Light on Dark Patterns. https://doi.org/10.2139/ssrn.3431205 (accessed 26/06/2021) (discussing the impact of dark patterns, including informational ones).

17 See e.g., O Katz & E Zamir, Do People Like Mandatory Rules? The Choice Between Disclosures, Defaults, and Mandatory Rules in Supplier-Customer Relationships, JELS (2021) 18(2) 421-60 (who compare the desirability of disclosures duties, from the perspective of the consumer, as compared to mandatory rules and default rules).

the DSA and DMA. Especially, we ask whether the meaning and use of terms that were discussed and finally became parts of information duties were fully shared among the stakeholders or not. For instance, terms like 'clear' or 'unambiguous' (referred to in Art. 24 DSA and extensively discussed before its adoption) are understood the same way by online platforms using personalized ads (addressed by the duty to disclose information) and the consumers (addressee of the information piece)? If this is not, could that be a source of disclosure ineffectiveness?

To assess if this is the case, we look at the stakeholder's submissions to the Commission's public consultation over three Inception Impact Assessment documents (IAs) that were input to the DSA and DMA proposals, namely: the so-called 'New Competition Tool',[18] the 'Ex ante regulatory instrument for large online platforms'[19] (hereafter also: ex ante tools), and the (then) 'Digital Services Act'.[20]

Second, we add computational analysis to standard manual reading of submissions that is done by the Commission without the help of algorithms.[21] The total of 2.862 replies to questionnaires and feedback documents contain the comments of all stakeholders regarding the proposals put forward by the Commission in its inception IAs. They, therefore, constitute an exceptional source of knowledge about who supported and opposed these duties among them, and especially, how individuals and organizations understand and use relevant terms of transparency. While manually processing the replies might still allow identifying the need for transparency duties, there are two shortcomings of this approach. First, any manual 'analysis' of the feedback documents comes with quite substantial labor cost, something that 'distant reading' can do more efficiently.[22] Second, no human reader can quantify the extent to which the same terms are used in the same way by different stakeholders. For instance, while both a large online platform and a consumer or smaller business might speak of a need for more 'precise' information, the underlying understanding and consequent use of this term could differ. In the context of transparency obligations, this is problematic since these duties might remain ineffective if a disclosure statement is only 'readable' in the eyes of the platform drafting it, but not in the eyes of the individual consumer or the micro organization reading it.

One way to cope with such limitations is to computationally analyze the feedback submitted to the Commission through the means of a mixed supervised and unsupervised ML technique, that would *complement* standard processing by public officials in the Directorates General (DG). Specifically, we propose doing so by using Word Embedding Alignment,[23] a state-of-the-art model for translation,[24] which can be adapted to our task, i.e. monolingual translation from a language to itself to evaluate the difference in the use of the same word in different corpora.[25] As a plus, word embedding modelling is highly compatible with unsupervised learning, a feature[26] that is very useful since, as explained before, in this context we should avoid the participation of human coding during the training process as much as possible.

This way, we aim to answer two central questions: (1) Do different groups of contributors share the same understanding (measured as semantical differences between terms) and use of the central terms and issues surrounding transparency and disclosure duties in the DSA and DMA? (2) Can we identify different clusters of opinions towards key concepts and can they be a possible source of disclosure failure? Our success in finding an answer to these questions with the help of said tools will be reflected with a view to a third overarching question: (3) can computational techniques help to partially automate the collection and analysis of opinions that are inputs to a rulemaking process? If this is the case, then we should recognize their potential in supporting the creation of better information disclosure rules, as is the proclaimed goal of the DSA and DMA consultation procedure, that is disclosure rules that are less prone to failure.

The article is structured as follows. The following section outlines the informational challenges posed by digital markets and the role of transparency duties set forth in the DSA and DMA proposals in mitigating their negative effects on consumers and businesses (Section 2). We then present our computational text analysis of the consultation documents and results, showing that not only are similar opinions expressed by groups that usually belong to different clusters (i.e., medium and big organizations); but also that groups of stakeholders use central terms in different ways (Section 3). We lastly conclude by sketching how a similar procedure could help to draft smarter disclosure regulations in a larger context.

2. Informational Malpractice in the Digital Era

For many commentators, the prominent role of transparency obligations in the DSA and DMA did not come as a surprise.[27] Disclosure

18 New Competition Tool, Inception impact assessment, Ares(2020)2877634, 4 June 2020, https://ec.europa.eu/info/law/better-regulation/have-your-say/initiatives/12416-New-competition-tool (accessed 31 March 2021).

19 The Ex ante regulatory instrument for large online platforms with significant network effects acting as gate-keepers in the European Union's internal market, Inception impact assessment, Ares(2020)2877647, 4 June 2020, https://ec.europa.eu/info/law/better-regulation/have-your-say/initiatives/12418-Digital-Services-Act-package-ex-ante-regulatory-instrument-of-very-large-online-platforms-acting-as-gatekeepers

20 The (then) Digital Services Act, Deepening the Internal Market and clarifying responsibilities for digital services, Inception impact assessment, Ares(2020)2877686, 4 June 2020, https://ec.europa.eu/info/law/better-regulation/have-your-say/initiatives/12417-Digital-Services-Act-deepening-the-internal-market-and-clarifying-responsibilities-for-digital-services_en.

21 R Senninger, Analyzing the EU Commission's Regulatory Scrutiny Board through quantitative text analysis. *Regulation & Governance,* (2020) 1; CM Radaelli, Regulating Rule-making via Impact Assessment. *Governance* (2010). 23(1), 89–108; CA Dunlop & CM Radaelli, Impact Assessment in the European Union: Lessons from a Research Project. *European Journal of Risk Regulation* (2015) 6(1), 27–34.

22 J. Grimmer & B.M. Stewart, Text as Data: The Promise and Pitfalls of Automatic Content Analysis Methods for Political Texts. *Political Analysis* (2013) 21(3), 267–297.

23 See e.g., D Alvarez-Melis & TS Jaakkola, Gromov-Wasserstein Alignment of Word Embedding Spaces. In *Proceedings of the 2018 Conference on Empirical Methods in Natural Language Processing*, pp. 1881–1890. Association for Computational Linguistics; Yehezkel Lubin, N., Goldberger, J., & Goldberg, Y. (2019). Aligning Vector-spaces with Noisy Supervised Lexicons. In Proceedings of the 2019 Conference of the North American Chapter of the Association for Computational Linguistics: Human Language Technologies, 460–465.

24 A Abdelsalam, O Bojar & S El-Beltagy, Bilingual Embeddings and Word Alignments for Translation Quality Estimation. *Proceedings of the First Conference on Machine Translation (2016): Volume 2, Shared Task Papers,* 764–771.

25 J Nyarko & S Sanga (2020). A Statistical Test for Legal Interpretation: Theory and Applications, 25 November 2020, https://juliannyarko.com/wp-content/uploads/other/nyarko_sanga_legal_interpretation.pdf. (showing how word embedding modelling can fit very well our task).

26 T Wada & T Iwata (2018). Unsupervised Cross-lingual Word Embedding by Multilingual Neural Language Models. arXiv:1809.02306 [cs]; A Conneau, G Lample, M Ranzato, L Denoyer & H Jégou, H. (2018). Word Translation Without Parallel Data. arXiv:1710.04087 [cs].

27 See e.g., Global Network Initiative (2020). Thinking Through Transparency and Accountability Commitments Under The Digital Services Act, 20 July 2020, https://medium.com/global-network-initiative-collection/thinking-through-transparency-and-accountability-commitments-un-

duties of all kinds have long been conceived as a key policy instrument to tackle the manifold challenges arising from digital markets. This section will give a snapshot of these challenges focusing and explaining the role of transparency in theory and in the DSA and DMA.

2.1 Talking at Cross Purposes. The Debate on the Need to Update Informational Duties through the DSA and DMA

Consumers benefit in many ways from the impressive development of digital markets.[28] However, certain characteristics of digital markets come with new challenges and risks. Concerning consumer protection, the sale of illicit goods in online marketplaces and unfair contractual clauses are key concerns.[29] But opaque online environments, as the Crémer report rightly emphasized, may also be 'a competition policy issue'.[30]

The relationship between transparency on the one side, and competition law and consumer protection, on the other, is bidirectional. A lack of competition might force business users to accept a level of transparency they do not feel comfortable with, in absence of an alternative supplier of the online service they are consuming.[31] This is an important realization since digital markets show certain characteristics which are likely to favor highly concentrated markets.[32]

Taken together, these factors work in favor of large online platforms, which might accumulate some kind of 'gatekeeping' power and impose the level of transparency they deem appropriate on the market they dominate. Of course, they technically still underly certain transparency obligations, for instance, those included in the GDPR.[33] However, the GDPR does not cover all relevant phenomena and users.[34]

Furthermore, platforms' understanding of specific requirements like e.g., 'clear and easy' language, might effectively determine the usefulness of disclosures for consumers, the small and medium enterprises. When consumers are not able to switch to a different provider giving information in a way that better fits their needs and capacities, a lack of competition could thus result in a lack of transparency.

The other way around, there are also situations in which a lack of transparency can endanger competition due to allowing for certain anti-competitive practices. In its investigation report on competition in digital markets, the US Congress subcommittee on Antitrust, Commercial Law and Administrative Law has summarized this as follows: 'Without transparency or effective choice, dominant firms may impose terms of service with weak privacy protections that are designed to restrict consumer choice, creating a race to the bottom'.[35] Clearly, that depends on the fact that in digital markets products are mainly zero-priced, and 'privacy and quality of service can be differentiating factors'[36]; hence, granting transparency or effective choice can help ensure competition.

Such a problem may arise in case platforms manipulate the order in which offers from business customers are presented.[37] Only if the parameters used to rank products are transparent, it will be possible to know whether an online platform is distorting competition by preferencing certain offers,[38] leaving consumers in the dark about the 'trade-offs they are facing', and hence inhibiting competition in a significant manner. In particular, self-preferencing by the big tech has been long debated as a cause of competition law infringement.[39]

der-the-digital-services-act-e4dce3cee909 (accessed 22 January 2021); S Stolton(2020). Make Big Tech accountable, Austria says in Digital Services Act recommendations, Euractiv, 30 November 2020, https://www.euractiv.com/section/digital/news/make-big-tech-accountable-austria-says-in-digital-services-act-recommendations/ (accessed 22 January 2021).

28 See Recital 1 DSA. To name just a few of these benefits: digital marketplaces facilitate cross-border trade and amplify product choices, social media allows cheap, easy, and quick communication, digital start-ups spur innovation and offer new services.

29 Concerning contractual clauses, an empirical analysis has identified potentially unfair contractual clauses in roughly 10% of a sample of 50 online consumer contracts. M Lippi, P Pałka, G Contissa, F Lagioia, H Micklitz, G Sartor & P Torroni, CLAUDETTE: An automated detector of potentially unfair clauses in online terms of service. *Artificial Intelligence and Law* (2019) 27(2), 117–139.

30 J Crémer, Y. de Montjoye & H Schweitzer, Competition policy for the digital era, European Commission Report (2019), https://data.europa.eu/doi/10.2763/407537 (accessed 14 February 2021) [hereinafter Crémer Report], 63.

31 This problem is well-framed as follows: 'a lack of options to switch to qualitatively similar other search engines or social networks might lead users to accept also very high prices (in form of collected data) and privacy policies that do not match their specific privacy preferences'. Kerber (n5) 867.

32 Crémer report (n 30) 2-3; M Gal & N Petit, Radical Restorative Remedies for Digital Markets. *Berkeley Technology Law Journal* (2020) 37(1), 5-6; OECD, Roundtable on Algorithms and Collusion - Executive Summary (DAF/COMP/M(2017)1/ANN3/FINAL), 26 September 2018, 5; F Scott Morton, P Bouvier, A Ezrachi, A Jullien, R Katz, G Kimmelman, D Melamed & J Morgenstern, Committee for the Study of Digital Platforms, Market Structure and Antitrust Subcommittee, Stigler Center for the Study of the Economy and the State [hereinafter Stigler report] (2019) 14. PG Picht & GT Loderer, Framing Algorithms: Competition Law and (Other) Regulatory Tools. *World Competition*, (2019) 42(3), 406.

33 Regulation (EU) 2016/679 of the European Parliament and the Council of 27 April 2016 on the protection of natural persons with regard to the processing of personal data and on the free movement of such data, and repealing Directive 95/46/EC (General Data Protection Regulation), 27 April 2016, O.J. L 119/1 [hereinafter GDPR].

34 For instance, the GDPR is not really relevant for business users, for it covers the personal data of individuals only (Art 2(1) in connection with Art 4(1) GDPR). It does not touch on the circumstances under which data (or content) deliberately shared by an individual can be removed by a platform. Neither does it regulate how data shared by a business user of an intermediary service should be displayed and what the user ought to know about this, which is central from a competition perspective.

35 U.S. House Committee on the Judiciary (2020). Investigation of Competition in Digital Markets. Washington, D.C.: Government Printing Office. The Subcommittee report also mentions manipulative design interfaces, so called dark patterns, nudging consumers into certain choices. Ibid, 53.

36 Ibid, 54.

37 Some authors argue that where consumer choices are being influenced, there is a special need for transparency duties: "A core element of such duties could be the obligation to thoroughly explain the workings of an algorithm, not on a technical level but regarding its impact on the customer, especially where it is designed to replace customer choice". Picht and Loderer (n 32) 416.

38 Contra, L Signoret, Code of competitive conduct: a new way to supplement EU competition law in addressing abuses of market power by digital giants. *European Competition Journal*, (2020). 16(2-3), 221, at 244 (contending that where platforms gain market power by being more efficient or winning consumers based on free choice by providing better offers, this would not constitute a violation of competition law).

39 Self-preferencing was at the heart of the Microsoft saga (see JP Jennings, Comparing the US and EU Microsoft Antitrust Prosecutions: How Level Is the Playing Field. *Erasmus Law and Economics Review*, (2006) 2, 71–86.) and was also heavily discussed by the doctrine at the time of the Google Shopping case. In fact, the Google Shopping case established that self-preferential placements are, indeed, not compatible with competition law. *Google Search (Shopping) Case C(2017) 4444*, 27 June 2017, paras 9, 10 of summary decision. See e.g., P Ackman, The Theory of Abuse in Google Search: A Positive and Normative Assessment Under EU Competition Law, in *Journal of Law, Technology & Policy*, (2) 301-372.

2.2 Legal Grounds for Updating Informational Duties

In the debate on how to react to some of these challenges, the e-Commerce Directive (ECD) has been central.[40] It is the piece of legislation the DSA updates and amends as 20 years of technological developments necessarily opened up some transparency-related lacunas.

First, platforms have quite simply become significantly larger and more important.[41] And with the reach of platforms, the amount of user-generated content has increased exponentially.[42] Hence, it is the increase in volume and magnitude of markets that justify a different approach. Second, existing rules were adopted when content moderation by automated means was not yet a widespread practice, if available at all.[43] Third, the increased relevance of recommender systems, digital nudging, personalized advertising also did not exist and was therefore not addressed by the ECD.[44]

Against the background of these developments, commentators and lawmakers have advocated in favor of significantly expanding the information duty framework of Arts 5, 6, and 10 ECD, with the aim of 'putting meaningful transparency at the heart' of new EU rules on digital services.[45]

With regards to the DMA, general shortcomings of EU competition rules when dealing with opaque online practices have been highlighted,[46] showing that law, albeit helpful, would most likely not suffice to achieve a satisfactory level of transparency.[47]

In light of these interconnected challenges for consumer protection and competition, the strong focus of the European Commission on informational duties as an easily enforceable means to increase transparency and mitigate information asymmetries seems reasonable in principle.[48]

However, over time, critics of information duties have continuously added evidence to the list of phenomena hampering the effectiveness of disclosures, which now includes e.g., information overload,[49] confirmation bias,[50] decision-making aversion,[51] the no-reading problem[52], and dislike.[53]

Despite this criticism, the Commission reports that 'many' in the consultation process have been calling for more informational duties. In the DMA, these 'many' correspond to civil society and media publishers, who 'called for an adequate degree of transparency in the market as well as the respect of consumers' autonomy and choice'.[54] In the DSA, the quest for 'algorithmic accountability and transparency audits, especially with regard to how information is prioritized and targeted' online comes from 'a wide category of stakeholders', and is particularly voiced by 'civil society and academics'.[55]

Apart from these brief notes, one cannot find more reference to the position of stakeholder groups with regards to transparency duties in the inception IAs. It is therefore relevant to see whether this synthesis duly captured the existing variegated positions. Before moving to our empirical analysis, we will briefly illustrate the actual transparency duties contained in the DSA and DMA proposals. These constitute the formalization of the debate we illustrated above, and we will use it as a blueprint for our empirical research.

2.3 The Actual Informational Duties in the DSA and the DMA

The European Commission's vision of what transparency rules might look like, as recently elucidated in the consultation on the DMA and DSA, will be briefly presented in the following. Some of these duties are new, while others are state-of-the-art for many operators. Indeed, especially those enlisted in the DSA are simply restated from the 2019 Platform-to-Business Regulation[56] and the amended Consumer Rights

40 Directive 2000/31/EC of the European Parliament and of the Council of 8 June 2000 on certain legal aspects of information society services, in particular electronic commerce, in the Internal Market (Directive on Electronic Commerce), 17 July 2000, O.J. L 178/1 [hereinafter ECD]; The ECD is considered by some as "the cornerstone of the Digital Single Market", European Parliament (n 8) 17.

41 Given that they reach a massive number of users, illegal or otherwise problematic content and practices will now impact considerably more citizens. SB Micova & A De Streel, Digital Services Act – Deepening the Internal Market and Clarifying Responsibilities for Digital Services, Centre on Regulation in Europe Report, 2 December 2020, https://cerre.eu/publications/digital-services-act-responsibility-platforms/ (accessed 16 February 2021) [hereinafter CERRE DSA Report], 10.

42 Alarmingly, this development has been associated with a rise in hate speech and disinformation. European Parliament, (n 8) 3.

43 Micova & De Streel (n 41) 10.

44 European Parliament (n 8), on page 12, mentions 'advertising, digital nudging and preferential treatment; paid advertisements or paid placement in a ranking of search results' as novel challenges to be addressed. Algorithm Watch (n 10) 1; European Parliament, (n 8) 5.

45 Algorithm Watch (n 10) 1.

46 The Crémer report points out several criticalities: (1) not all gatekeepers enjoy a dominant position in the sense of Art. 102 TFEU; (2) the relevant market might be substantially harder to define than in non-digital cases; (3) not every problematic practice has a demonstrable effect on the relevant market. The authors conclude that greater emphasis should be put on the theory of harm, instead. Crémer report (n 31) 3-4. Moreover, digital markets are often moving at a rapid pace, which is not necessarily a characteristic they share with competition law. Hence, there are concerns whether competition law could be applied with the necessary speed to address urgent competition needs. A de Streel, Digital Markets Act – Marking Economic Regulation of Platforms Fit for the Digital Age, Centre on Regulation in Europe Report, 24 November 2020 [hereinafter CERRE DMA report], 59; Recital 5 DMA.

47 Information duties have also increasingly been acknowledged as competition remedies by courts, partly shifting from traditional cease and desist orders towards transparency duties see SW Waller, Access and Information Remedies in High-Tech Antitrust, *Journal of Competition Law and Economics* (2012) 8(3), 575, at 576.

48 JC Coffee, Market Failure and the Economic Case for a Mandatory Disclosure System. *Virginia Law Review* (1984) 70(4), 717–753; SJ Grossman & JE Stiglitz, Information and Competitive Price Systems. *The American Economic Review* (1976) 66(2), 246–253; SJ Grossman & JE Stiglitz, On the Impossibility of Informationally Efficient Markets. *The American Economic Review* (1980) 70(3), 393–408; PG Mahoney, Mandatory Disclosure as a Solution to Agency Problems. *The University of Chicago Law Review* (1995) 62(3), 1047–1112.

49 HA Simon, A Behavioral Model of Rational Choice. *The Quarterly Journal of Economics* (1955) 69(1), 99–118.

50 A Tversky & D Kahneman, Judgment under Uncertainty: Heuristics and Biases. *Science* 1(1974) 185(4157), p. 1124–1131.

51 O Ben-Shahar & CE Schneider, The Failure Of Mandated Disclosure. *University of Pennsylvania Law Review* (2011) 159, 727, Ildd (2015) (nt 11).

52 For an empirical investigation of this issue, see Y Bakos, F Marotta-Wurgler & DR Trossen, Does Anyone Read the Fine Print? Consumer Attention to Standard-Form Contracts. *The Journal of Legal Studies*, (2014) 43(1), 1–35.

53 Katz & Zamir (n 17).

54 DMA, at 8 (summarizing the results of stakeholder consultations and impact assessments).

55 DSA at 9. See also Algorithm Watch (n 10) 1; CERRE DSA report (n 41) 39; European Parliament, (n 8), 5; European Commission, White Paper on Artificial Intelligence - A European approach to excellence and trust, COM/2020/65 final, 19.2.2020, 15.

56 Regulation (EU) 2019/1150 of the European Parliament and of the Council of 20 June 2019 on promoting fairness and transparency for business users of online intermediation services OJ L 186, 11.7.2019, p. 57–79.

Directive[57].

2.3.1 DSA: Arts. 12(1), 13, 23-25, 29 and 33

As summarized in Table 1 in the Appendix, the DSA proposal includes a variety of transparency and disclosure obligations (together: informational duties) for providers of intermediary services.[58]

Art 12(1) would entail a general obligation to inform users about potential restrictions to their services contained in the terms and conditions. This information would need to be publicly available, provided in an *easily accessible format*, and written in *clear and unambiguous language.*

Whereas agreeing to the terms and conditions of a platform can be a one-time action, Art 13 DSA would oblige platforms to publish yearly reports about their content moderation practices. These reports would need to be drafted in a *clear and comprehensible language* and include certain specific information.[59]

While these obligations would apply to all providers of intermediary services, online platforms would additionally have to provide information about the out-of-court dispute settlements, content suspensions, and the use of automatic tools for content moderation (Art 23 DSA). Concerning the latter, the platform would be obliged to elucidate the '*precise* purposes, indicators of the accuracy of the automated means in fulfilling those purposes and any safeguards applied'. Consequently, it seems fair to expect that the understanding of terms like 'precise' 'clear' 'unambiguous' would be crucial factors in determining the scope and form of the information provided to users.[60]

For online platforms displaying advertisements, Art 24 DSA would establish further informational duties. Advertisements and their publishers would have to be identifiable in a 'clear' and 'unambiguous manner'. Furthermore, platforms would have to share 'meaningful information about the main parameters used to determine the recipient to whom the advertisement is displayed' with the platform user. In addition to the obligations laid down in Art 24 DSA, very large online platforms within the meaning of Art 25 DSA,[61] would further need to offer application programming interfaces (APIs) to access information on the advertisements they display (Art 30(1), (2) DSA).

Apart from advertisement algorithms, rankings and recommender systems have been identified above as another platform architecture component requiring increased transparency.[62] For very large online platforms this challenge is addressed by Art 29 DSA: in their terms and conditions, very large online platforms would have to flag the use of recommender systems and explain in a 'clear, accessible, and easily comprehensible manner' how these systems work (i.e., which parameters they use and how they can be modified or influenced).[63] Again, the question of how simple, precise and understandable disclosures are understood seems central regarding the *de facto* effect of these transparency duties.

Lastly, Art 33 sets out comprehensive transparency obligations for very large online platforms.[64] These more pronounced transparency obligations for very large online platforms reflect the differentiated approach the Commission took for the design of the DSA, explicitly mentioned in Recital 39 of the proposal.[65]

2.3.2 DMA: Arts 5(g) and 6(1)g

The bottom part of Table 1 clearly shows that transparency duties in the DMA are more scarce than in the DSA and mostly relate to rankings and advertising services.[66] They are nonetheless a breakthrough in competition law, because they are ex ante policies envisaged to prevent severe hindrance to market forces from occurring. That justifies the choice to analyze them here.

The main provisions of interest are Arts 5(g) and 6(1)g DMA, especially if read in combination with Recitals 42 and 53. Art 5(g) DMA would oblige gatekeepers, with respect to their core platform services (within the meaning of Art 3(7) DMA), to 'provide advertisers and publishers ..., upon their request, with information concerning the price paid by the advertiser and publisher, as well as the amount or remuneration paid to the publisher'.[67]

Furthermore, advertisers and publishers can request, and obtain free of charge access to performance measuring tools and the information that is needed to perform their own verification to assess how satisfied they are with the advertisement product they are paying for (Art 6(1)g DMA).

While these obligations are rather specific, Art 10 DMA would open the door to add further transparency duties in the future if a market investigation pursuant to Art 17 DMA identified a need to do so for the sake of safeguarding fair competition.

57 Directive (EU) 2019/2161 of the European Parliament and of the Council of 27 November 2019 amending Council Directive 93/13/EEC and Directives 98/6/EC, 2005/29/EC and 2011/83/EU of the European Parliament and of the Council as regards the better enforcement and modernization of Union consumer protection rules OJ L 328, 18.12.2019, p. 7–28.

58 Above (n 7). In Table 1 (Appendix), we specify whether the norm imposes a transparency or disclosure obligation. Here we use the two as synonyms.

59 i.e., the number of removal orders received from Member States, categorized by the type of illegal content and the average time required to remove such content; the amount of notice submitted pursuant to Art 14, any action taken thereupon, average time needed for this action, own-initiative, content moderation measures affecting availability, visibility and accessibility of information, and the number of complaints received by the internal complaint system (Art 17 DSA).

60 For a discussion of the 'clearly, comprehensibly, and unambiguously' requirement in Art 10 e-Commerce Directive, see A Lodder & A Murray, EU Regulation of E-Commerce. (Edward Elgar Publishing 2017), 26. While case law on the matter is rather sparse, the ECJ clarified that information that can only be accessed by a number of clicks is still provided in a clear and comprehensible manner. *Bundesverband der Verbraucherzentralen und Verbraucherverbände - Verbraucherzentrale Bundesverband eV v Amazon EU Sàrl, Case C649/17*, 10 July 2019, para. 52.

61 Per the thresholds chosen by the Commission for the designation of very large online platforms under Art 25(2) DSA and the relation with the different notion of gatekeeper in the DMA see nn 3 and 6 above.

62 Recital 62 DSA.

63 Moreover, the service recipient would have to be provided with an easily accessible functionality allowing her to select her preferred option for the recommender system the platform is using (Art 27(2) DSA).

64 Not only do they have to publish reports every six months (instead of yearly), they also have to include a risk assessment (pursuant to Art 26 DSA), risk mitigation measures (pursuant to Art 27 DSA), audit reports (pursuant to Art 28(3) DSA), and audit implementation reports (pursuant to Art 28(4) DSA).

65 For a thorough discussion of how differentiating rules better ensure the proportionality of regulatory intervention, see F Di Porto & N Rangone, Behavioural Sciences in Practice: Lessons for EU Policymakers. In A Alemanno and A Sibony (eds) Nudge and the Law, (Hart pub 2014) 20-59. With reference to transparency duties, Di Porto and Maggiolino (n 12) 12-22. See also CERRE DSA report (n 41) 11.

66 Note that we are focusing on general informational duties, not those which only apply if there is an investigation underway (see Art 19 DMA).

67 This is a self-enforcing obligation for gatekeepers vis-à-vis advertisers and publishers to which they provide advertising services. Gatekeepers should inform about the price paid their counterparts as well as the remuneration paid to the publisher for the publishing of an ad and for the advertising services provider by the same gatekeeper. Such transparency duty, as clarified in Recital 42, is needed for the parties to better understand the real value of the service provided.

To sum up, this section has shown that despite the many criticisms, transparency duties loom large in the DSA and DMA proposals. By analyzing in greater detail the actual disclosure duties of the two acts, we provided evidence of the way the Commission seeks to attain a high level of consumer protection and fair competition for digital services.

The analysis shows a stark contrast between what most commentators critique regarding the utility to enact more transparency duties and what the proposals purport. That suggests exploring other and new research routes to understand how these duties were implemented in the DSA and DMA proposals.

3. A Computational Analysis of The DSA and DMA Consultation Process

In this section, we ask whether informational duties are what stakeholders asked for in the consultation process and whether their actual wording in the DSA and DMA reflects the way each group uses the relevant terms. This is a relevant step, as it is important that those who implement disclosure duties (typically digital firms, be they small, medium or large) and the beneficiaries of information (individuals, but also micro-organizations) agree on the meaning of the duties (e.g., 'clear', 'accessible', or 'unambiguous language').

To do so, we leverage the power of ML and computational text analysis techniques. In the following, we present our empirical analysis of the replies and position papers submitted by stakeholders to the EU consultation process for three inception IAs. We first give a high-level description of our methodology (for a more detailed description, see Appendix),[68] before presenting our results.

3.1 Our Methodology

We collected and analyzed a total of 2,862 replies to the questionnaires and 1,862 of the respective feedback documents attached to the replies.[69] In total, we built a dataset of 3,032,418 words. To do so, we automatically downloaded all the relevant files from the Commission's website.[70] Unlike the replies (in excel), most attached submissions came in PDF format, so we first converted them into text and then constructed three large clusters.

3.1.1 Groups Identification

To identify groups of stakeholders, we relied on the Commission's categorization scheme for the organization 'size' of the feedback contributors, which groups feedback comments from (1) individuals, micro (10 employees), (2) small (50 employees), (3) medium (250 employees), and (4) large (250 or more) organizations.[71] We then aggregated the different sub-categories (3) and (4) to form three larger categories:

A. individuals and micro firms/organizations;

B. small firms/organizations; and

C. medium and big firms/organizations.

As explained in the previous paragraph, the initial clusters were based on European Commission's 'size' division. From that clustering, we aggregated medium and big firms, as suggested by: (1) the cluster size, and (2) a Kolmogorov-Smirnov test performed on the questionnaires accompanying the consultation (further explained in the Appendix).

Neither the size of companies nor the questionnaire answers we chose to perform the K-S test on were re-used for the Word Embedding Modeling (see below, A.2), hence avoiding double-dipping.

Our decision on *how* to do this aggregation was based on a qualitative and quantitative analysis of the questionnaire accompanying the feedback documents.[72]

This allowed us to find out which groups of consultation participants are the most similar and should be clustered together. Note that a Kolmogorov-Smirnov test we performed on the categorical (i.e., multiple-choice) questions in the questionnaire showed that 'medium and large' entities should be grouped together as they can be assumed to be one cluster.[73] This is per se a relevant finding, because although different in size, and despite the fact that in most economic surveys they are considered separately, medium and large entities are a cluster for the purpose of text analysis. That is justified by both qualitative and quantitative factors.

First, our algorithm assessed replies provided by firms *and* organizations together, while in economic surveys just *firms* are grouped in one cluster. It is therefore possible that the presence of organizations attenuated the distance in the use of terms.

Second, that is extremely relevant because even if medium and large entities decide through different mechanisms (e.g., taking a decision may involve only one manager in medium organizations, while requiring dozens in big ones), what we assess is the way they understand and use terms related to transparency duties. Hence, the size of

68 The methods we used and describe hereafter largely overlap with those described in F. Di Porto et al., I see something you don't see. A computational analysis of the DSA and the DMA, appeared in (2021) Stanford Computational Antitrust, (1)6. However, there we focused our analysis on terms related to competition in digital markets and used the theoretical legal framework typical of antitrust law. In this paper, we deploy algorithms on informational duties proposed by the DSA and DMA and use theories of regulation to interpret the results of our computational analysis.

69 Note that the replies were used partially: we only employed those drafted in English and related with disclosure terms (we manually coded these: see Appendix for further details).

70 All the documents we used can be found under the following links. As per the DSA proposal: European Commission, Digital Services Act – deepening the internal market and clarifying responsibilities for digital services, 11 January 2021, https://ec.europa.eu/info/law/better-regulation/have-your-say/initiatives/12417-Digital-Services-Act-deepening-the-Internal-Market-and-clarifying-responsibilities-for-digital-services (accessed 28 January 2021) As per what became the DMA proposal: European Commission, Digital Services Act package – ex ante regulatory instrument of very large online platforms acting as gatekeepers, 11 January 2021, https://ec.europa.eu/info/law/better-regulation/have-your-say/initiatives/12418-Digital-Services-Act-package-ex-ante-regulatory-instrument-of-very-large-online-platforms-acting-as-gatekeepers; and European Commission, Single Market – new complementary tool to strengthen competition enforcement, 11 January 2021, https://ec.europa.eu/info/law/better-regulation/have-your-say/initiatives/12416-New-competition-tool.

71 The Commission distinguishes the feedback also by 'types' of contributors. E.g respondents to the DSA were: the general public (66%), companies/businesses organizations (7.4%), business associations (6%), and NGOs (5.6%) authorities (2.2%), academic/research institutions (1.2%), trade unions (0.9%), and consumer/environmental organizations (0.4%) (see DSA at 8).

72 See European Commission (n 57) for the questionnaire. A detailed description of how we analyzed the questionnaire can be found in the Appendix.

73 This choice can not only be backed by our data, but also by some scholarly findings, e.g., R Kemp & C Lutz,Perceived barriers to entry: Are there any differences between small, medium-sized and large companies *International Journal of Entrepreneurship and Small Business,* (2006) *3(5), 538–553*. For a more detailed description as to why we cumulated medium and large entities, instead of clustering medium with small ones, see the Appendix.

organizations is not a relevant parameter, as it is semantics.

Third, by analyzing the text of organizations' opinions, as formalized in the feedback documents and replies, and later encapsulated in the DMA and DSA informational rules, we are able to capture how medium and large entities make use of terms related to transparency.

3.1.2 Word Embedding Modelling: Training the Algorithm

After having identified the most sensible way to cluster the consultation documents, we built three corpora:

- 744 documents with 35,949 unique words for corpus A (*Individuals and micro enterprises and organizations*),
- *393 documents with 32,100 unique words for corpus B (small companies/organizations),*
- *and 689 documents with 39,815 unique words for corpus C (medium and large companies/organizations).*

We always compared two corpora, hence we analyzed three corpus pairs (A-B, B-C, A-C).

By constructing three different corpora, we were able to train a neural network on the documents of each cluster, hence having three networks that capture the intricacies of each corpus. Based on the number of times words occur next to each other, this network allowed us to calculate a vector for each word in each corpus, a so-called Word Embedding Model (more specifically, we used Gensim's CBOW word2vec model).[74] These models are remarkable in the sense that they can capture the semantic meaning of words in a set of numbers. For instance, in a well-trained model, the distance between the vector of the words 'Paris' and 'France' will be roughly the same as between 'Rome' and 'Italy'. Hence, the relative positions of vectors in the model approximately represent the meaning of certain terms. This means that while a simple algorithm would require researchers to formulate explicit rules to approximate the semantic meanings of words, ML (or the neural network, to be precise) learns the implicit rules directly from the data we feed it. This does not only increase the performance of the algorithm but also prevents an undue influence of the researchers' conscious or subconscious assumptions.[75]

3.1.3 Making sense of semantic distance

However, it needs to be noted that models trained on different corpora are not directly comparable. Since the vectors making up the models are based on the frequency of words occurring next to each other, they depend on the corpus the model was trained on. Hence, even the position of words that most definitely have the same meaning for all groups (e.g., 'and') will have very different vectors, which we would normally interpret as a semantic difference. In this case, however, the distance between the two vectors will not be the result of a different use of a word, but simply the particularities of the corpuses the model was trained on. Consequently, to make the models we trained on the different corpuses comparable, we used unsupervised vector space alignment. This allowed us to bring the vectors trained on two different corpuses together in one model space, where they would be comparable. Put differently, in the aligned model space, strongly differing vectors represent actual differences in the use of a word, instead of being a result of a different training basis.

However, we still needed to ascertain that these differences were not merely incidental, but actually of a certain significance. To do so, we employed a statistical test. This test relies on the assumption that the distance between the vectors for the same word from two different corpora can be split into three components: a semantic difference (i.e., a difference in meaning), a non-semantic difference (e.g., syntactical differences), and a random difference. We then set two assumptions: first, we assume that the semantic difference between corpora for a certain set of words (the control vocabulary) is zero. This means that we assume all stakeholder groups use words like 'and' or 'one' in the same way. Based on this, we were able to construct an empirical distribution of the non-semantic difference and the random difference, assuming that there is no semantic difference. This distribution is our second assumption.

Knowing how our vectors should look like if there was no semantic difference between the clusters, we were then able to check for each word if the distance between its vectors from two different corpora is compatible with this hypothesis of a uniform use. If it is not, we can conclude with a certain level of confidence that there is a statistically significant difference in its semantic meaning between the different corpora.

With these tools at hand, we analyzed the stakeholder submissions to the DSA and DMA consultation process. Given that the stakeholders whose opinions we analyze are to a large extent those who will either draft or receive the abundant transparency statements envisioned in the proposals,[76] their uses and view of terms related to informational duties should be of great interest both for legislators and scholars debating the factual role of informational obligations.

The questionnaires raise several points, not all of which immediately related to informational duties. For instance, the NCT questionnaire also discusses competition problems (such as agreements, self-preferencing, or collusion); while the DSA one includes questions on liability of intermediaries.

Because we are interested in the use of certain terms only, we created an initial list of 119 terms, based on the glossaries of the consultation questionnaires which explain terms that might be new to some consultation participants. However, after the first analysis, we realized that our list of terms might be too narrow for two reasons.

First, the wording of the Inception Impact Assessments (IIAs) which were discussed in the consultations differs from the final draft DSA and DMA. The change in vocabulary is especially marked in the DMA,[77] where classic concepts of competition law (such as market, dominance, efficiency gains) are mostly abandoned, and new ones are defined.[78] Since we used corpora from comments to the three IIAs

74 T Mikolov, K Chen, G Corrado & J Dean, (2013). Efficient Estimation of Word Representations in Vector Space. ArXiv:1301.3781 [Cs]. R eh ek, R. (2019). Word2vec embeddings. https://radimrehurek.com/gensim/models/word2vec.html (accessed 22/06/2021).

75 For instance, a researcher might assume that a word needs to be used in the same sentence at least *x times for the two to be related and design her algorithm accordingly. For our algorithm, we do not need these kinds of assumptions or rules as the algorithm learns directly from the data.*

76 This includes the general public, authorities and consumer/environmental organizations (as addressees), and companies/businesses organizations, business associations, and trade unions (as drafters); but will exclude NGOs, and individual academics and research institutions.

77 The difference in terminology also derives from the fact that the 'NCT' inception IA was based on Art 106 TFEU (much focused on competition), while the 'Ex-ante regulation's legal base was Art 114 TFEU (internal market). Following the consultation, the DMA proposal had its own legal base (Art 114) and terminology.

78 As are spheres of application of the DMA in comparison with the inception IAs.

documents to run our analysis and needed it to reflect this change, we proceeded with hand-coding. Therefore, we combined words from two sources: (i) all glossaries[79] attached to previous legislation (all EU Directives and Regulations) that were recalled by the DSA and DMA proposals (for a total of 119 words); and (ii) terms related to transparency (e.g. 'disclos*', 'transparency', 'inform*' and the like) that were manually selected from the questionnaires (102 words). As a result, we ended up with a list of 194 words (102 from the DSA's questionnaire and 92 from the DMA's). (See Annex 3.1).

Furthermore, since we are interested in the specific provisions of the DSA and DMA which qualify how information should be provided (e.g., 'clear', 'accessible'), we added all those terms from the proposals' informational provisions (ten terms in total, see Annex 3.1).

Finally, stakeholders use a variety of terms to refer to the same concept. For instance, our list might include 'self-preferencing', but we would miss differences on 'self-favoring'. Our pre-defined list of terms was not able to capture this variety. Since it was also not feasible to anticipate all these variations, we chose to manually code those results that are closely related to the terms and concepts of our list *ex post*.

79 Glossaries are definitions of terms usually contained in Arts. 2 of EU Directives and Regulations. Namely, we added all the glossaries from: the GDPR, the NIS Directive, the Data Governance Act proposal, the E-commerce Directive and the Platform-2-Business directive.

To perform manual coding, we relied on the legal expertise of our team, with the aid of external assistance.[80] Finally, the terms that were added manually were a total of 204, while overall the computational analysis was performed over of a total of 323 words.

3.2 Results: Different Groups, Different Uses?

We found a statistically significant difference for

1,865 word pairs between corpora A and C,

2,184 between corpora A and B and

1,113 between B and C.[81]

A detailed description of how this comparison was conducted and what 'significant' means in this context, is provided for in the Appendix (Annex 3). From all the 5,162 significant distances we found, we chose those that were relevant to our analysis, based on the selection procedure described above. This resulted in a list of 13 relevant terms

80 We are thankful to Andrea Ruffo, legal scholar and teaching assistant at Luiss University of Rome for his wonderful assistance in the manual coding activities. The legal analysis was performed by Tatjana Grote and Fabiana Di Porto.

81 It needs to be noted that many of these words are not of particular interest for us because they might identify a specific service of a certain company (e.g., the 'Gmail' email service in Google's submissions). However, some of the key buzzwords surrounding competition and transparency obligations show statistically significant differences.

Table 1: Summary of results

Term	Distance AB	Distance BC	Distance AC	Close words A	Close words B	Close words C
Consumer-centric	1.557 (0.03)**	1.625 (0.02)**	1.247 (0.16)	privacy-protecting	systems	computing
Easy	1.444 (0.04)**	1.443 (0.07)*	1.451 (0.05)*			
Easy-to-use	1.450 (0.04)**	1.427 (0.08)*	1.522 (0.02)**	deregulation		cut-off
Meaningful	0.545 (0.627)	0.670 (0.648)	1.482 (0.04)**			
Precise	1.645 (0.01)**	0.878 (0.434)	0.747 (0.497)	cartel	checklist	
Privacy-friendly			1.468 (0.04)**	misconceptions		tailor-made
Ranking	1.182 (0.15)	1.644 (0.02)**	1.452 (0.05)*		guidelines, improve, oversight	appearance, disclosing
Readable	1.051 (0.237)	1.720 (0.01)**	1.394 (0.08)*		effective, specific, clear	entities
Self-regulatory	1.340 (0.09)*	1.536 (0.04)**	0.897 (0.37)		blacklisting, sanctions, obligations	benchmarking, codes, ameliorate
Simple	1.703 (0.01)**	1.504 (0.05)*	1.158 (0.20)	formats	precise	
Understandability		1.663 (0.02)**			single-homing, practice	informs
Unregulated	1.361 (0.07)*	1.566 (0.04)**	1.822 (0.00)***	not-sufficient		mitigation
Well-informed	0.943 (0.293)	1.734 (0.01)**	1.749 (0.00)***	Confusing, explainable		Inscrutability, implementation

Note: The asterisks indicate significance at a 0.001 (***), 0.05 (**), and 0.1 (*) level, respectively.

for which we found significant differences in use and understanding.

Table 1 shows these results. The 'Distance' columns report the distance between the vectors of the same words for each corpus pair, with the respective p-value in parentheses. A grey field in the 'Distance' columns indicates that a word was not used in both of the respective corpora.

The 'Close Words' columns shine a light on some of the concepts that were closely related with the term in question in the corpora for which there was a statistically significant distance between the terms. To be precise, we computed the ten words which were most similar to the term in question[82] and then hand-coded those words which were relevant to our analysis, based on the same procedure outlined above (see the last paragraph of 3.1.2). A grey field in the 'Close words' columns means that we did not look for close words because the respective corpus was not involved in any of the significant distances or there were no meaningful close words.

Moving on to our results, we start with some terms that are of importance on a meta-level, namely those related to the overall regulatory strategy employed. Since there are different regulatory paths to ensuring transparency (e.g. by regulation or self-regulation), this is of interest as well.[83]

3.2.1 Words related to the regulatory 'meta-level'

We observe that '**self-regulatory**' is used differently by different stakeholders. Generally, we see that self-regulation seems to be a more prominent issue for medium and big companies (corpus C): while the term is only mentioned ca. 5,000 times by small companies (corpus B, with 810, 961),[84] it occurs more than 25,000 times in corpus C (which contains 1,177,120), where it is associated with the terms 'benchmarking', 'codes', and 'ameliorate'. This is reflected in Fig. 1, and could be read as a sign that self-regulation is seen as an important strategy by medium and big companies/organizations.

Differences in use also exist for the term 'unregulated'. For individuals (A) and small entities (B), an 'unregulated' digital single market does not seem like a favorable option, with 'not-sufficient' and 'precariousness' as closely related terms. (Fig. 2)

3.2.2 Words related to informational duties

With regards to *informational duties*, it is interesting to note that there is a statistically significant distance between the use of the word '**simple**' between corpus A and B (Fig. 2). While individuals and micro-businesses/organizations seem to focus on 'formats' regarding simplicity, small companies/organizations in our dataset associate the attribute '**precise**'. However, it needs to be noted that the term 'precise' also underlies some significant differences between corpora A and B, which is an important finding in light of the wording of Art. 23 DSA (Table 1).

Generally, individuals and micro-organizations (A) used the word '**simple**' roughly 20-times more often than small businesses and organizations (B).

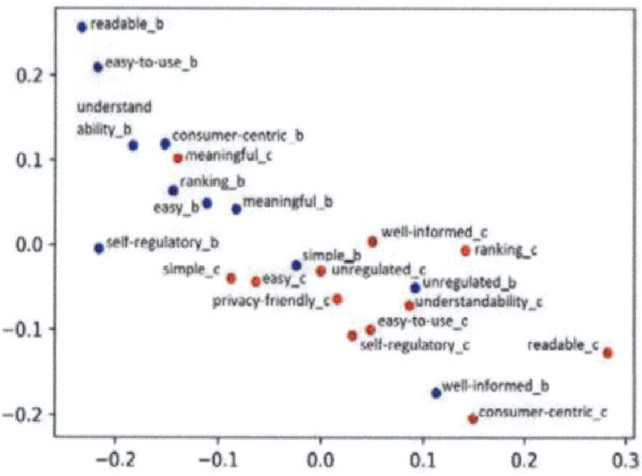

Figure 1: Aligned Vector Space Model - Corpora B & C

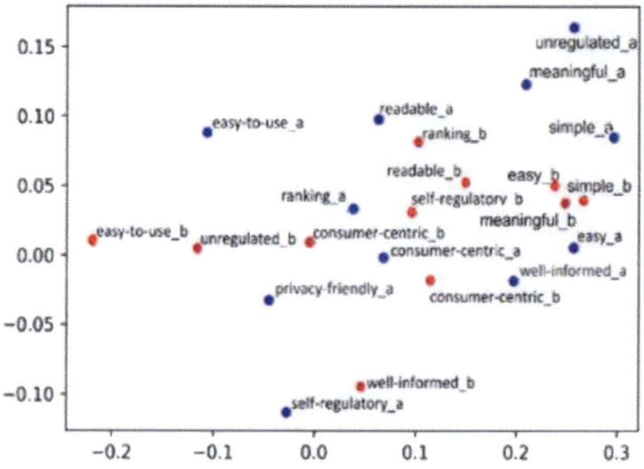

Figure 2: Aligned Vector Space Model - Corpora A & B

With regards to the obligation of advertisement system transparency laid down in Art 24 DSA, it is surprising to see that '**meaningful**' is used very differently by individuals and micro-organizations/businesses (A) than by medium and big companies (C).[85] Again, this could potentially impact the efficacy of said provision since what is deemed 'meaningful' by the drafters of the respective disclosures might be rather meaningless for their recipients.

In the comparison between corpora A and C, the term '**well-informed**' is mentioned roughly 26,000 times by individuals and micro-contributors (A; in total 1,044,337 words) compared to 18,642 mentions in corpus C (in total 1,177,120 words) and is closely related to 'explainable'. Furthermore, we find a different utilization of the terms '**easy-to-use**' and '**privacy-friendly**', respectively (see Fig. 3).

The first is interesting with a view to rules like Art 17(2) DSA, which speaks of *easy to access, user-friendly* complaint mechanisms. The latter seems to be located within slightly different contexts by different stakeholder groups: while individuals (A) heed possible 'misconceptions', medium and large companies/organizations (C) associate '**privacy-friendly**' with 'tailor-made' and 'reinforced'. Interestingly, the Commission explicitly mentions that '**privacy-friendly services**' were

82 Our similarity measure is the cosine distance between two vectors. R eh ek, R. (2019). Gensim: Store and query word vectors - Similarity. https://radimrehurek.com/gensim/models/keyedvectors.html#gensim.models.keyedvectors.WordEmbeddingsKeyedVectors.similarity (accessed 30/08/2020).

83 For a detailed discussion of regulatory strategies in disclosure regulation, see Di Porto & Zuppetta (n 14).

84 Note that the corpus sizes indicated here refer to the overall corpus, i.e., the number of words in the documents as they were submitted. For corpus sizes indicated above we only considered the unique words for each corpus, which is why these numbers are much smaller.

85 The use of 'Meaningful' for the corpus pair C and A might look close in Fig. 3 because the difference is not as pronounced as for some other terms, but it has p-value of 0.04, meaning that we can conclude there is a statistically significant difference.

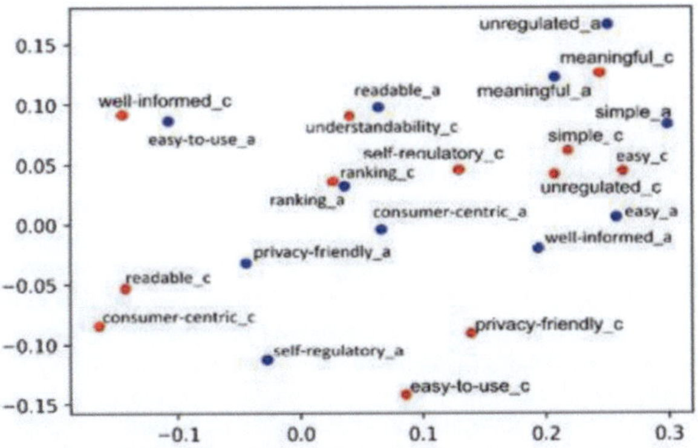

Figure 3: Aligned Vector Space Model - Corpuses A & C

one key expected outcome of the DMA in the eyes of the consultation respondents. However, what might be missing is that not all stakeholders understand the same when speaking of 'privacy-friendly'.

Comparing small companies/organizations (B) and medium/big companies/organizations (C), we find a significant distance between the vectors for the terms '**well-informed**' and '**consumer-centric**' (Fig. 1, above). The latter word is closely related to the term 'systems' in corpus B, which is unsurprising. In corpus C, we see a close association with 'computing', which is interesting since it seems to shift the focus of consumer-centric design to the processes happening behind the systems that consumers interact with.

Another intricate finding concerns the term '**ranking**', which has been central in discussions about the transparency of online platforms. This close connection between transparency and rankings is also reflected in the close words we found: small companies (B) associate rankings with *'guidelines'*, medium/big companies with *'disclosing'*.

As 'ranking' is not a crucial term for transparency duties as such, this difference will not necessarily impede the effectiveness of disclosures. Nevertheless, this finding shows that there are different perceptions of some key concepts of the DSA and DMA across stakeholders.

We further find differences for the terms '**understandability**' and '**readable**'. This should be a key concern for policymakers and legal scholars when debating transparency duties: if no uniform understanding of what 'readable' transparency disclosures look like can be reached, consumers will likely have to deal with strongly differing levels of readability and understandability.

3.3 Challenges

Our algorithmic analysis of the consultation process for the DSA and DMA has shown that there are statistically significant differences between stakeholders' use and understandings of some key concepts of transparency. To the best of our knowledge, we are the first to conduct such a 'close reading' of an EU rulemaking process and discern differences in the ways a consultation relates to the rules in the context of the DSA and DMA. Our results show that NLP techniques can allow the Commission to understand not only what stakeholders say, but what they actually mean; which could substantially improve stakeholder consultations' analysis as we did here. For instance, the Commission took note of demands for more 'simple' notice-and-action procedures for content removal.[86] Yet, we discovered that the term 'simple' might not be understood in the same way across all groups of stakeholders. This could offer a first signal to the Commission that it is premature to legislate on this matter; or that a one-size-fits-all measure may not be suitable.

Linking our results back to the discussion of transparency duties and their importance for consumer protection in digital markets, our findings cast doubt on whether all stakeholders have a similar understanding and thus make similar uses of **simple**, **meaningful**, **easy-to-understand**, **readable** transparency statements. Given that the exact implementation of such duties often lies in the hands of different stakeholders, this might be one reason why transparency duties remain ineffective. For instance, our algorithm reveals that 'meaningful' is understood and used differently by the individual consumers and the medium/big platforms. This may cause Art. 24 DSA failure, as it obliges platforms to inform consumers in real-time that what is being displayed to them is an ad, in a clear and '*unambiguous* manner'. Since the literature on the failure of disclosure regulation has mostly focused on how transparency statements are perceived by consumers,[87] our focus on all stakeholders, inclusive both the recipients and drafters of disclosure statements, adds a unique, novel perspective.

Having said that, there are challenges that need to be addressed, some of which are common to the computational law scholarship,[88] others are specific to our analysis. Both offer room for improvement by future research.[89]

Concerning the analysis, in the methodology, we make two assumptions for the statistical test we perform: that words in the control vocabulary used for the vector space alignment transformation do not have a semantic difference and that the distribution of distances has the same shape also for the other words. For instance, we assume that words like 'and' or 'one' are understood in the same way by all contributors in the consultation. While this seems plausible, we cannot entirely discard the possibility of errors in the creation of the models and their alignment due to shortcomings in these assumptions. Nonetheless, our assumptions are commonly accepted in the literature.[90]

Second, our corpora are relatively small and heterogeneous since they contain documents from many different authors with potentially different styles and focuses. For instance, feedback we analyzed are in English language only, but their authors might not be native English speakers. This could introduce a bias, meaning that results may be partially driven by the particularities of our corpora. Hence, increasing the corpus size and the control vocabulary should be a top priority for future research. Another way to solve the problem would be using bootstrapping: by repeatedly and randomly changing some words in the corpora and then taking the mean value, the random term u_t^{AB} in the distribution of distances could be reduced.

Generally, it needs to be noted that our analysis focuses on the *identification* of semantically different terms. At this stage, we do not seek to provide insights into what the identified differences might be based on and how they impact the stakeholders' opinions. Therefore, it has some limitations as far as *interpretation* is concerned. Using word embedding alignment alone does not allow (yet) to show any causal relationship between differences in perceptions of transpar-

86 DSA proposal, 8.

87 Above (n 11).

88 D Lim, Can Computational Antitrust Succeed? *Stanford Computational Antitrust*, https://law.stanford.edu/wp-content/uploads/2021/04/lim-computational-antitrust-project.pdf (accessed 22/06/2021), 10-13.

89 More technical limitations are presented in the Appendix.

90 See Nyarko and Sanga (n 25), 4.

ency and specific factors. Although we compared the most similar vectors[91] corresponding to the word pairs of interest, gaining an idea of how the meanings might differ, this still requires a certain degree of *ad hoc* interpretation. Moreover, we used ex post manual coding when selecting the results to be presented here. In the future, fully replicable, *ex ante* criteria should be used to make this selection.

Due to these limitations, our results need to be treated with caution and should be complemented by further research. Nevertheless, they constitute a first step providing interesting insights into informational duties in the DMA and DSA.

4. Concluding Remarks

This paper sets out to explore whether different stakeholders participating in the consultation process for the latest Commission proposals on new rules for digital markets (the DSA and DMA) share a similar understanding of key concepts related to one integral pillar of the new proposals: informational duties. We analyzed the replies to questionnaires and feedback documents submitted in the consultation process using the NLP technique of Word Embedding Alignment, which allowed us to identify terms that are not used in the same way by all stakeholders.

We find significant differences in the way stakeholders use words that are central in transparency duties, like 'readable', 'simple', and 'privacy-friendly'. These differences are group-specific, and hold for individuals and micro organizations; small; and medium/large organizations. If that might seem obvious at first sight, it is surprising if one considers that those participating in the consultation process on the DSA and DMA constitute a rather small epistemic community, made of legal and economic scholars, digital companies, NGOs, and IP specialists who have a high stake interest in expressing their voice and are, therefore, well-informed about the subject they discuss.

Our results should be a key concern for policymakers and legal scholars for several reasons. Differences in understanding might mean (undesirable) differences in implementation. If there is no uniform use (and understanding) of what 'readable' transparency disclosures or 'simple' complaint mechanisms look like, users will likely have to deal with strongly differing levels of readability and simplicity.

Second, this could decrease the effectiveness of transparency duties in ensuring competitive and fair markets, given that those who replied to the consultation are also those who will draft and receive the disclosures.

Third, and strictly related, different understanding and uses of words that are relevant to informational duties might also help explain why such rules fail.

The last takeaway we want to stress is that rule-makers are recommended to consider another interesting finding: that understanding and use of relevant terms of transparency (like 'simple' and 'well-informed') do not differ between medium and big organizations (corpus C), as one would expect. That is to the point to make them a sole group for the sake of text analysis. Generally, if the Commission used tools like the one applied here to complement its impact assessments and rulemaking, it could not only hear what stakeholders *say* but understand what they *mean*, which might ultimately improve the functioning of the EU's new regulatory traffic lights for digital markets.

Looking at the perspectives this paper opens, we think that our analysis, if complemented with other computational techniques, will be very useful in doctrinal studies of the future.

One scenario could be to investigate the 'rationale' of the DSA and DMA's rules. By the time the DSA and DMA will entry into force, their wording will change several times, depending on multiple interactions of the Commission, the Parliament, Council and stakeholders. Our analysis might be a first step in the direction of keeping records of textual modifications and then tracing back the statements that influenced them the most (e.g., being the most similar). Clearly, our analysis alone would not be enough and would need to be complemented with other NLP techniques. For example, text similarity techniques could be employed to map out which stakeholder opinions might have influenced the EU institutions when drafting not only its proposals but also its final rules. This might allow gaining a precise understanding of why rules were drafted in a certain way and could greatly help the interpretation of rules in light of their *telos* and their drafting history.

A second research area that our analysis could inaugurate is that of improving the drafting of disclosure statements and transparency reports, as envisaged by the two new proposals. While we considered the use and understanding of information-related terms by firms and organizations together, one could zoom in on the use of concepts by individual consumers and firms only, which will certainly differ. For instance, the phrase 'easy to use' was used differently by all three clusters. If we already find this disagreement in large, aggregated groups, the understanding of such a phrase will most likely differ between individuals. Consequently, regulators might opt for clusterized disclosures, with messages adapted to the specific informational capabilities of users' groups (as identified by our computational analysis).

That might help to overcome many of the shortcomings of current disclosure statements. While this possibility was discussed in great detail elsewhere,[92] our analysis suggests that the Commission and platforms would be well-advised to explore this possibility.

Our algorithm should be seen as the first building block of a fully-fledged tool for a more in-depth algorithmic analysis of EU rulemaking. The other building blocks might be:

- 'topic modeling',[93] which would allow rule-makers like the Commission and scholars to get an intuitive understanding of how the most important topics, that will become rules in a near future, are part of a shared view among different stakeholders or whether they emphasize different issues;
- 'document similarity'[94] could be used to cluster statements that are input to regulation before the Commission publishes a regulatory proposal. This could help to perceive certain similarities or alliances, between stakeholders, even across different groups like

91 See n 82 above.

92 See, e.g., F Di Porto, Algorithmic Disclosure Rules, in Artificial Intelligence and Law, (2020), https://ssrn.com/abstract=3705967 or http://dx.doi.org/10.2139/ssrn.3705967 (accessed 27 October 2021). More information on the implementation of clusterized disclosures is available at: www.lawandtechnology.it. See also: Busch, C. (2019). Implementing Personalized Law: Personalized Disclosures in Consumer Law and Data Privacy Law. *The University of Chicago Law Review* 86(2), 309–332.

93 DM Blei, AY Ng & MI Jordan, Latent dirichlet allocation. *The Journal of Machine Learning Research*, (2003) 3, 993–1022.

94 See, e.g., BK Triwijoyo & K Kartarina, Analysis of Document Clustering based on Cosine Similarity and K-Main Algorithms. *Journal of Information Systems and Informatics*, (2019) *1(2), 164–177*. DG annemann, Comparative Law: Study of Similarities or Differences?, in M. Reimann and R. Zimmermann (eds.), *Oxford Handbook of Comparative Law* (2d ed.) (Oxford University Press, 2019).

e.g., small companies and medium/large companies.

- Sentiment Analysis could be another means to understand if the parties to a rulemaking process agree or disagree with certain proposals or statements. In fact, we performed a first explorative sentiment analysis using a pre-trained model on those paragraphs in our documents which contain the terms of interest presented above (Table 1). While this analysis produced some interesting results,[95] a fully-developed sentiment analysis is best left for future research. Furthermore, one could cluster each statement based on the overall sentiment of a group of contributors[96] to get a better understanding of how supporters and critics of a proposal are distributed and what their main concerns and arguments are.

Overall, while we believe that discerning latent differences in the use of certain terms is a crucial capability that could significantly enhance the consultation process at the EU level, the above-mentioned additions could be combined in a fully-fledged NLP toolbox that could substantially enrich the work of both the Commission and legal scholars and provide many new insights.

Be that as it may, it is hoped that our findings will enrich the positive and normative debate about transparency rules in digital markets, inspire future research in the computational antitrust arena, and urge EU rule-makers to rethink their convictions about the use of computational tools in the consultations.

Addendum

Corrigendum - The authors also published a paper using the same dataset and methodology in Fabiana Di Porto, Tatjana Grote, Gabriele Volpi & Riccardo Invernizzi, "I see something you don't see": A computational analysis of the Digital Services Act and the Digital Markets Act", 2021 *Stanford Computational Antitrust journal, #5* https://law.stanford.edu/wp-content/uploads/2021/08/di-porto-computational-antitrust.pdf.

95 For instance, we found that 'understandability' is seen much more favorably by small companies/organizations (B; 0.721) than by medium/big entities (C; 0.340). Similarly, we found a more positive attitude towards the terms 'well-informed' and 'consumer-centric' for individual and micro contributors (0.624) than for small companies/organizations (0.051). We also identified a negative sentiment of small companies/organizations towards the term 'unregulated' (-0.118). Lastly, 'simple' is viewed more favorably by individuals and micro contributors (A; 0.314) than by big and medium organizations/businesses (C; 0.220).

96 See e.g., S Feng, D Wang, G Yu, C Yang & N Yang, Sentiment Clustering: A Novel Method to Explore in the Blogosphere. In Q Li, L Feng, J Pei, SX Wang, X Zhou, & QM Zhu (Eds.), *Advances in Data and Web Management.* (Springer 2009) 332–344.

Appendix

Table 1 Informational duties in the DMA and DSA

T / D duty	Digital Services Act (DSA)	Recipient of info (r) Info to be provided (i)	'How' to disclose	Core service providers (Art 2(f) DSA)	Online platforms (Art 2(h) DSA)	Very Large online platforms (Art 25)
D	Terms of service include information on content moderation and use of algorithms	(r) Users; (i) potential restrictions to their services.	'easily accessible format' written in 'clear unambiguous language'	Art 12 (Terms and conditions)		
T	Yearly reports on content moderation providing key information specified in Art 13(1) DSA	(r) Users and the general public; (i) content moderation practices	written in 'clear and comprehensible language'; need to include specific information (a. 14, 17)	Art 13 (Transparency reporting obligations for providers of intermediary services)		
D	Reasons for removing the content or disabling access	(r) Users whose content was removed or access disabled	Clear and specific statement containing the information listed in Art 15(2)		Art 15 (Statement of reasons)	
T	Additional information (with reference to Art. 13) on content suspension actions taken, use of automated means for content moderation, and out-of-court dispute settlement	(r) Users and the general public, (i) esp. about automation of content moderation and ADR	Format potentially to be specified by Commission, Art 23(4)		Art 23 (Transparency reporting obligations for providers of online platforms)	
T/D	Advertising transparency duties	(r) Users and recipients of service; (i) display that info is an ad + personalization of ad	Provided in a 'clear and unambiguous manner'		Art 24 (Online advertising transparency)	
D	Main parameters used in recommender systems must be set out in terms and conditions	(r) Users; (i) use of algorithms for recommending content	Provided in a clear, accessible, and easily comprehensible manner		Art 29 (Recommender Systems)	
T	Additional advertisement transparency duties to maintain in the repository and made accessible	(r) Users and the general public; (i) advertisements and their display	Repository be made publicly available through an API	/	Art 30 (Additional online advertising transparency)	
T	Additional information on content moderation, risk management, and auditing	(r) Users, the general public, and Digital Service Coordinator; (i) results of risk assessments and audits	-		Art 33 (Transparency reporting obligations)	
	Digital Markets Act (DMA)	**Recipient of info**	**'How' to disclose**	**Gatekeepers** **(as defined in Art 3 DMA)**		
D	Information about advertising services provided by gatekeepers for advertisers and publishers	(r) Advertisers and publishers counter-parts	-	Art 5(g) (Obligations for gatekeepers)		
D	Provide free of charge access to performance measuring tools of gatekeepers and information necessary to enable advertisers to carry our independent verification	(r) Advertisers and publishers	-	Art 6(g) (Obligations for gatekeepers susceptible of being further specified)		

Note: Informational duties (Column 1) may include either transparency duties (T) or disclosure duties (D).

Annex 1 Groups identification

To analyze the replies to questionnaires and feedback documents, we created a special scraper algorithm, which allowed us to download all the files automatically, convert them into text, and split them into three clusters. In doing this, we started by following the Commission's categorization scheme for the organization size of the feedback contributors. We then aggregated the different sub-categories into three corpora based on the typology and the dimension of the feedback contributor: Corpus A (individuals and micro organizations), B (small companies/organizations), and C (medium and large companies/organizations).

Our clustering choice is based on two considerations: First, a qualitative analysis of the questionnaires accompanying the feedback documents[97] allowed us to get an understanding of which aggregation would cluster comparable feedback contributors together. We mostly analyzed the types of feedback contributors in the sample and had a look at their replies to questions related to informational duties. Second, we conducted a quantitative analysis of the same questionnaires to ensure that our clusterization choices are solid. In particular, we sought to ensure that there is no statistically significant difference between medium and large entities in our sample since at least medium companies are often grouped with small, rather than large companies.[98] However, it needs to be noted that our feedback contributors are not only businesses but also other types of organizations. This diversity could "smooth" the differences we would have expected to find if our sample included companies only. In fact, our qualitative analysis of the questionnaires suggested that medium entities in our sample are more comparable to large businesses/organizations both in terms of entity type (whether they are from academia, civil society, private economy, etc.) and in terms of how they perceive challenges arising from digital markets (in the sense that they gave more similar answers to the pertinent multiple-choice questions in the questionnaires).[99] To test the robustness of this perception, we analyzed the answers provided for by medium and large entities to specific multiple choices questions.[100] We applied a Kolmogorov-Smirnov two-sample test[101] to understand if there is a statistically significant discrepancy between the distribution of the answers of the two groups. If that was the case, we would assume that these answers must be considered as provided by two different populations, not allowing us to treat them as a unique cluster. The results of the test are shown in Figure 1.

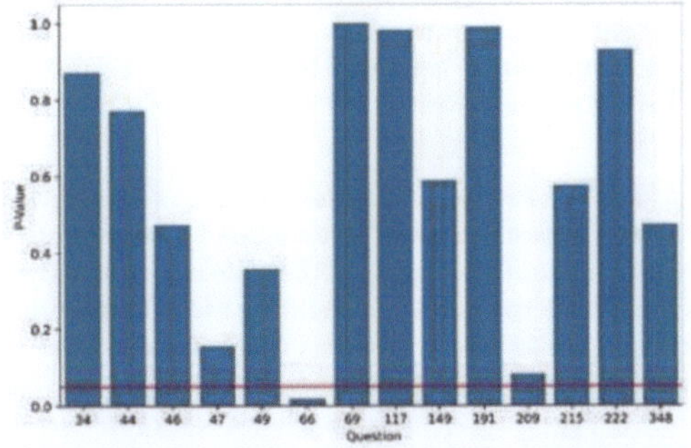

Figure 1: p-values resulting from KS-two sample test applied to the answer distributions of the considered questions. Red line highlights our significative tolerance value of p=0.05

Even using a very high tolerance **p-value** level of 0.05, only question no. 66 showed a statistically significant variation. This question alone however is mostly unrelated to our core research interest, and hence unlikely to compromise the validity of our clustering.

In total, we collected 744 documents with 35.949 words for corpus A, 393 documents with 32.100 words for corpus B, and 689 documents with 39.815 words for corpus C. We always compared two corpora, hence we analyzed three corpus pairs (A-B, B-C, A-C).

Annex 2 Training the algorithm

To discern differences in the use of certain key terms across stakeholder groups (i.e., a different semantic understanding of identical terms), we leveraged Word Embedding Models to quantify evidence of such differing understandings. This technique has already been used in various Natural Language Processing tasks, and recently also in the Computational Law literature.[102] It has been demonstrated to be very powerful and useful in providing insights into latent differences in how language is used.

The core of this technique consists in training a special neural network to convert each word contained in a corpus of texts into a vector, i.e., a set of numbers.[103] While a simple algorithm would require researchers to formulate explicit rules to somehow approximate the semantic meanings of words, ML (or the neural network, to be precise) learns the implicit rules directly from the data we feed it. This does not only increase the performance of the algorithm but also

97 European Commission, Digital Services Act – deepening the internal market and clarifying responsibilities for digital services, 11 January 2021, https://ec.europa.eu/info/law/better-regulation/have-your-say/initiatives/12418-Digital-Services-Act-package-ex-ante-regulatory-instrument-of-very-large-online-platforms-acting-as-gatekeepers/public-consultation (accessed 28 January 2021).

98 Statistically significant refers to the hypothesis of the K-S test, that the data of both groups is originating from the same population.

99 While this could be due to the idiosyncrasy of our sample, this finding also corresponds with scholarly literature. See e.g., R Kemp & C Lutz. Perceived barriers to entry: Are there any differences between small, medium-sized and large companies. *International Journal of Entrepreneurship and Small Business*, 2006 3(5), 538–553.

100 The questions were selected manually based on two criteria: First, we manually identified all questions relating to informational duties and competition in digital markets. In a second step, we singled out questions that had a categorical answer scale, i.e., non-text replies.

101 L Hoboes Jr. The significance probability of the Smirnov two-sample test. *Matematica* 1958 3(5), 469-486.

102 See e.g., Nyarko and Sanga (n 25); E Peramo, C Cheng & M Cordel, Juris2vec: Building Word Embeddings from Philippine Jurisprudence. *2021 International Conference on Artificial Intelligence in Information and Communication (ICAIIC)*, 121–125; I Chalkidis & D Kampas, Deep learning in law: Early adaptation and legal word embeddings trained on large corpora. *Artificial Intelligence and Law*, 2019 27(2), 171–198; A Mandal, K Ghosh, S Ghosh, S & S Mandal, Unsupervised approaches for measuring textual similarity between legal court case reports. *Artificial Intelligence and Law*, 2021 29(1):1-35.

103 The Neural Network in particular is a LSTM (Long-Short Term Memory Network). See S Hochreiter & J Schmidhuber, Long Short-term Memory. *Neural Computation* 1997 9(8):1735-80. More generally, see S Lai, K Liu, S He & J Zhao, How to Generate a Good Word Embedding. *IEEE Intelligent Systems*, 2016 31(6), 5–14; Y Li & T Yang, Word Embedding for Understanding Natural Language: A Survey. In S. Srinivasan (Eds.), *Guide to Big Data Applications*. Springer International Publishing, 2018 83–104.

prevents an undue influence of the researchers' conscious or subconscious assumptions. The resulting vectors are based on the frequency of words occurring next to each other, meaning their relative positions in each phrase of the corpus and the correlation between words. The stronger two words are correlated (in their occurrence – and so in their semantic meaning)[104] in the corpus the model was trained in, the closer the corresponding vectors will be located to each other.

However, the meaning of the vectors in the model depends on their relative positions in the respective corpus; the vector of a single word alone does not give us any insights. To test if there is evidence of different semantic use of the same words between two texts, we had to assess the distance between vectors from the two different corpora corresponding to the same words. To align them, we transformed the two models geometrically.[105] This allows us to understand how a vector in one corpus relates to the vector of another corpus. After the transformation, the vectors of the two aligned corpora are comparable to each other.

For each corpus we trained a different word embedded space, and we aligned each pair of words occurring in both corpora through the means of Unsupervised Vector Space Alignment.[106]

Annex 3 Making sense of semantic distance

3.1 The Data

1. List of terms from glossaries[107]

E-commerce directive, P2B regulation, glossary of terms for DSA' questionnaire:

1. Application Programming Interface
2. Collaborative Economy Platform
3. Competent Authorities
4. Content Provider
5. Digital Service
6. Harmful Behaviours
7. Activities Online
8. Hosting Service Provider
9. Information Society Service
10. Illegal Content
11. Illegal Goods
12. Illegal Hate Speech
13. Intermediary Service
14. Intermediation Services
15. Law Enforcement Authorities
16. Notice
17. Notice Provider
18. Online Advertising
19. Online Platforms
20. Online Platform Ecosystems
21. Recommender Systems
22. Scaleup, Smart Contracts
23. Start-up
24. Trusted Flagger
25. User
26. Gatekeeper
27. Core Platform Service
28. Digital Sector
29. Online Intermediation Services
30. Online Search Engine
31. Online Social Networking Service
32. Video-Sharing Platform Service
33. Number-Independent Interpersonal Communications Service
34. Operating System
35. Cloud Computing Services
36. Software Application Stores
37. Software Application
38. Ancillary Service
39. Identification Service
40. End User
41. Business User
42. Ranking, Data
43. Personal Data
44. Non-Personal Data
45. Undertaking
46. Control
47. Recipient
48. Consumer
49. Offer Services
50. Trader
51. Intermediary Service
52. Illegal Content
53. Dissemination
54. Distance Contract
55. Online Interface
56. Digital Services Coordinator Of Establishment
57. Digital Services Coordinator Of Destination
58. Advertisement, Recommender System
59. Content Moderation
60. Terms And Conditions
61. Service Provider
62. Established Service Provider
63. Commercial Communication
64. Regulated Profession
65. Coordinated Field
66. Business User
67. Provider
68. Corporate Website User
69. Ranking
70. Mediation
71. Durable Medium

104 This is based on the 'distributional hypothesis', which assum es that words which frequently occur together are usually also semantically related. While this approach might seem too simple to capture complex semantic meanings, the success of algorithms relying on it suggests that the claim has some merit. E Altszyler, M Sigman, S Ribeiro & DF Slezak, Comparative study of LSA vs Word2vec embeddings in small corpora: A case study in dreams database. *Consciousness and Cognition* 2017 56, 178–187.

105 To perform this transformation, we used a "control vocabulary", containing a list of words that we can safely assume that share the same semantical meaning . The list of 1,189 words we used is, in fact, composed mainly of numbers and stop-words (like e.g., 'the'). We are thankful to Professor Julian Nyarko from Stanford University for providing us with a first list of Control keywords, to which we further added almost 2000 numerals and stop-words from the different corpuses.

106 We used a special algorithm provided by Facebook in the library FastText. (https://github.com/facebookresearch/fastText), used in Python. P Bojanowski, E Grave, A Joulin, & T Mikolov,. Enriching Word Vectors with Subword Information, 2017. http://arxiv.org/abs/1607.04606 (accessed 22 January 2021).

107 Terms gathered from glossaries attached to all legislation recalled by the DSA and DMA proposals plus terms taken from the glossary attached to the DSA questionnaire.

From DGA proposal:

72. Access
73. Re-Use
74. Metadata
75. Data Altruism
76. Data User
77. Data Holder
78. Data Sharing Main Establishment
79. Public Sector Body
80. Bodies Governed by Public Law
81. Public Undertaking
82. Secure Processing Environment
83. Representative

From NIS (Network and Information Systems):[108]

84. Network And Information System
85. Security Of Network And Information Systems
86. National Strategy On The Security Of Network And Information Systems
87. Operator Of Essential Services
88. Digital Service Provider
89. Incident
90. Incident Handling
91. Risk
92. Standard
93. Specification
94. Internet Exchange Point (IXP)
95. Domain Name System (DNS)
96. DNS Service Provider
97. Top-Level Domain Name Registry
98. Online Marketplace

From GDPR:

99. Processing
100. Restriction Of Processing
101. Profiling
102. Pseudonymisation
103. Filing System
104. Controller
105. Processor
106. Third Party
107. Consent
108. Personal Data Breach
109. Genetic Data
110. Biometric Data
111. Data Concerning Health
112. Enterprise
113. Group Of Undertakings
114. Binding Corporate Rules
115. Supervisory Authority
116. Supervisory Authority Concerned
117. Cross-Border Processing
118. Relevant And Reasoned Objection
119. International Organisation

108 EU rules on the security of Network and Information Systems (NIS) are at the core of the Single Market for cybersecurity. The Commission proposes to reform these rules under a revised NIS Directive to increase the level of cyber resilience of all relevant sectors, public and private, that perform an important function for the economy and society. https://eur-lex.europa.eu/legal-content/EN/TXT/PDF/?uri=CELEX-:32016L1148&from=EN

II. Manually coded from the questionnaires on DSA and DMA

Manually coded from Questionnaire for the public consultation on a New Competition Tool

1. Access to data
2. adjacent/neighbouring markets
3. aftermarket
4. algorithm-based technological solutions
5. alignment of prices
6. anti-competitive
7. appropriateness
8. barriers to enter
9. binding
10. case-by-case
11. choice
12. competition
13. concentrated market
14. conditions of competition
15. copyright
16. customer lock-in
17. customer switching costs
18. data accumulation
19. data dependency
20. digital markets
21. digitisation
22. dominance-based
23. dominant
24. dual role situations
25. economies of scale
26. economies of scope
27. extreme economies of scale
28. fixed operating costs
29. gatekeeper
30. global distribution footprint
31. homogeneity of products
32. incomplete or misleading information
33. increased transparency
34. incumbency advantages
35. incumbency advantages
36. information asymmetry
37. innovation
38. inspections
39. interim measures
40. investigative powers
41. judicial review
42. lack of access to data
43. lack of competition
44. lack of transparency
45. leveraging
46. lock-in effects
47. market concentration
48. market dominance
49. market entry
50. market player
51. market power
52. market share
53. market-sharing cartels
54. monopolisation
55. multi-homing
56. multi-sided markets

57. network effects
58. new competition tool
59. non-binding recommendation
60. oligopolist
61. oligopolistic market structures
62. oligopoly
63. online platform
64. patents
65. penalties
66. platform
67. policy options
68. price increases
69. price leader
70. price leader-follower behavior/behaviour
71. price-fixing
72. pricing algorithms
73. procedural safeguards
74. proportionality
75. recommendations
76. regulatory barriers
77. related market
78. request of information
79. single-home
80. start-up costs
81. structural lack of competition problem
82. structural risk for competition
83. switching
84. tacit collusion
85. tailored remedies
86. tipping
87. tipping markets
88. transparency
89. two-sided markets
90. vertical integration
91. voluntary commitments
92. zero-pricing

Terms manually coded from **DSA questionnaire**

1. accountability
2. advertisement
3. algorithmic process
4. app store
5. appropriate
6. auction
7. automated detection
8. banning
9. bargaining power
10. behavioural advertising
11. blog hosting
12. bullying
13. business users
14. child sexual abuse material
15. complaint
16. conglomerate
17. conglomerate effect
18. consumer rights
19. content moderation
20. contestable
21. contextual advertising
22. control mechanism
23. counter-notice
24. coverage
25. cyber security
26. data sharing
27. dependency
28. digital identity
29. disabling
30. discrimination
31. disinformation
32. disputes
33. dissemination
34. divisive messages
35. due diligence
36. effective
37. effective measures
38. enforcement
39. ex-ante rules
40. fast-track assessment
41. flagging
42. fundamental rights
43. gender equality
44. governance
45. grooming
46. harmful
47. hate speech
48. illegal content
49. illegal medicine
50. information disclosure
51. institutional cooperation
52. internal practices
53. interoperability
54. know your customer
55. large online platform companies
56. leverage
57. liability
58. manipulation
59. market entry
60. national level
61. non-discrimination
62. non-payment
63. notice-and-action
64. notice-and-takedown
65. notifications
66. operating systems
67. oversight
68. pet trafficking
69. platforms' content policies
70. political advertising
71. price comparison
72. primary activities
73. programmatic advertising
74. proportionate
75. quality standards
76. Rating and reviews
77. Real-time bidding
78. recommendation
79. redress
80. Referral
81. reinstated content
82. removal
83. remuneration
84. reporting procedure

85. search engines
86. sector specific rules
87. self-employed
88. sharing
89. social networks
90. solidarity
91. suspension
92. tailored
93. takedowns
94. terrorist propaganda
95. trusted organisations
96. trusted researchers
97. unfair
98. unfair practices
99. unfavorable
100. user base
101. very large online platform companies
102. video sharing

Terms manually coded from the DSA and DMA proposals:

1. easily accessible
2. clear
3. unambiguous
4. specific
5. easily comprehensible
6. available
7. detailed
8. easy to access
9. user-friendly
10. precise

3.2 Statistical test

To see if there is evidence for a statistically significant semantic difference between the use of a term between the different stakeholder groups, we must perform a statistical test of their relative distance. We can model the relative distance d_t^{AB} dABt of a word t in the corpus A and B be as:

$$d_t^{AB} = \gamma_t^{AB} + \mu_t^{AB} + u_t^{AB}$$

This takes into account a semantical term γ_t^{AB}, a non-semantical term (originated from the simple different words disposition in the two corpora) and a random term . More precisely, the semantic term is defined as the difference in the usage of the same word which is driven by different understandings of the meaning of this term. Hence, this is the term we are interested in. On the other hand, the non-semantic term is defined as the term capturing all the non-semantic differences in usage, which can emanate from more frequent use of the word in different contexts, different authors, or stylistic differences. Finally, we define the random term as random differences in usage unrelated to systematic differences between the corpora. These could arise from the document-production process or the randomness of the initialization of the word-embedding algorithm's training.[109]

The statistical test we performed is based on two assumptions. Our first assumption is that words in the control vocabulary used for the Vector Space Alignment Transformation do not have a semantic difference, i.e., $\gamma_t^{AB}=0$. Consequently, their relative distance can give an empirical distribution of the non-semantical distance between words, composed of the only two terms $\mu_t^{AB} + u_t^{AB}$ which is our second assumption. In this manner, it is possible to construct an empirical cumulative distribution of these distances, distributed with the hypothesis of zero semantic difference.

We first built an empirical Fisher-Snedecor distribution of distances calculated with all the common words included in the Control Vocabulary. We then analyzed the distance between the vectors of a word in the two corpora, counting the number of times these values were smaller than the control words' distances in the distribution. If we accept the null hypothesis that the word we are analyzing shows no semantic difference between the different corpora, then the obtained (normalized) **p-value** tells us the probability to have a distance equal or greater than that. If this probability is small enough, we can refuse this null hypothesis with a small possibility of error. This is to say that the particular word has, indeed, a statistically significant semantic difference in the two corpora. A general acceptance value for the p-value is 0.05, which we will use as the critical threshold for our analysis.

Annex 4. Cumulative distribution of semantic differences

Figures 1 to 3 show the cumulative distribution of distances of control dictionary words (in blue) against the cumulative distribution of distances and similarities of analyzed words (in red) for each corpus pair (i.e., corpus X against corpus Y). The plot shows that the words we analyzed create a statistical distribution different from the one of the common words, as we can see from the different shapes. These differences suggest that there are significant semantic differences between the corpora.

Figure 1. Corpuses AB - Cumulative distribution of control distances (top) and similarities (bottom)

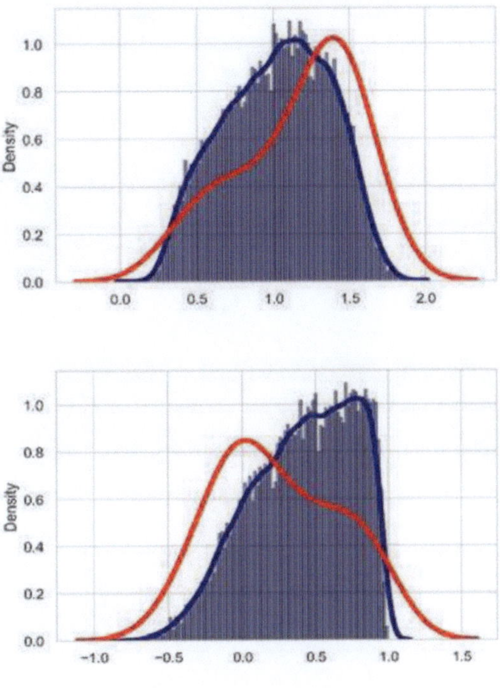

109 Nyarko & Sanga (n 102).

Figure 2. Corpuses BC - Cumulative distribution of control distances (left) and similarities (right)

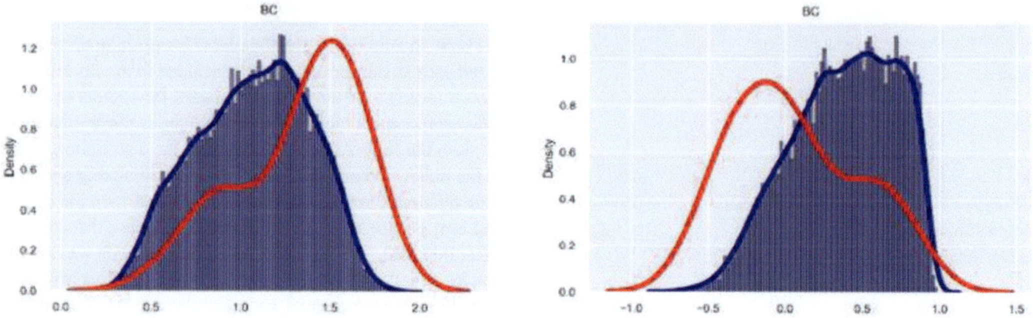

Figure 3. Corpus Pair AC - Cumulative distribution of control distances (left) and similarities (right)

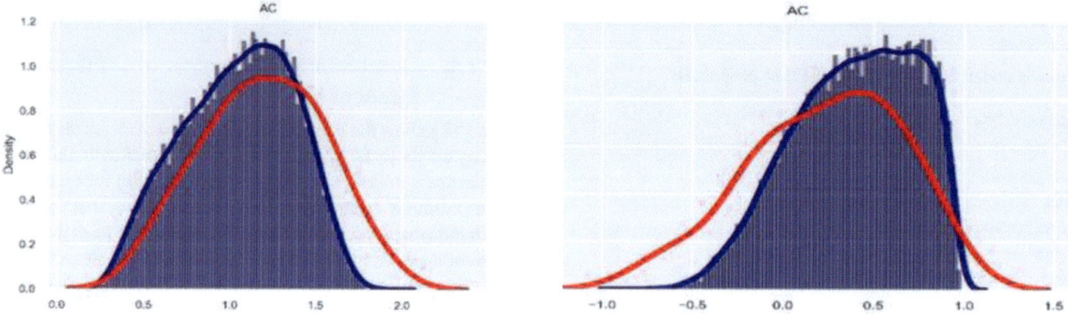